Date D

D0436527

About Island Press

ISLAND PRESS, a nonprofit organization, publishes, markets, and distributes the most advanced thinking on the conservation of our natural resources—books about soil, land, water, forests, wildlife, and hazardous and toxic wastes. These books are practical tools used by public officials, business and industry leaders, natural resource managers, and concerned citizens working to solve both local and global resource problems.

Founded in 1978, Island Press reorganized in 1984 to meet the increasing demand for substantive books on all resource-related issues. Island Press publishes and distributes under its own imprint and offers these services to other nonprofit organizations.

Support for Island Press is provided by Apple Computers Inc., Mary Reynolds Babcock Foundation, Geraldine R. Dodge Foundation, The Charles Engelhard Foundation, The Ford Foundation, Glen Eagles Foundation, The George Gund Foundation, William and Flora Hewlett Foundation, The Joyce Foundation, The John D. and Catherine T. MacArthur Foundation, The Andrew W. Mellon Foundation, The Joyce Mertz-Gilmore Foundation, The New-Land Foundation, The J. N. Pew, Jr., Charitable Trust, Alida Rockefeller, The Rockefeller Brothers Fund, The Florence and John Schumann Foundation, The Tides Foundation, and individual donors.

The Living Ocean

DISCARD

The Living

DISCARD

$\mathbb{O}$cean *Understanding and Protecting Marine Biodiversity*

Boyce Thorne-Miller and John G. Catena

Foreword by Sylvia Earle

The Oceanic Society of Friends of the Earth, U.S.

ISLAND PRESS

Washington, D.C. □ Covelo, California

574.52636
T 511 L

© 1991 Friends of the Earth

All rights reserved. No part of this book may be reproduced in any form or by any means without permission in writing from the publisher: Island Press, Suite 300, 1718 Connecticut Avenue NW, Washington, D.C. 20009.

The author gives grateful acknowledgment to Alice Jane Lippson for the use of her illustrations, which appeared originally in *Life in the Chesapeake Bay*, 1984: Drawings © 1984 by Alice Jane Lippson.

Library of Congress Cataloging-in-Publication Data

Thorne-Miller, Boyce L.
   The living ocean : understanding and protecting marine biodiversity / Boyce L. Thorne-Miller and John G. Catena ; foreword by Sylvia Earle.
      p.   cm. — (Island Press critical issues series ; #4)
   Includes bibliographical references and index.
   ISBN 1-55963-064-7
   1. Marine biology. 2. Biological diversity. I. Catena, John G. II. Title. III. Series.
   QH91.T49   1991
   574.5′2636—dc20   90-44024
                                                                      CIP

Printed on recycled, acid-free paper

Manufactured in the United States of America

10 9 8 7 6 5 4 3 2

1/92

Contents ▄▄▄▄▄▄▄▄▄▄▄▄▄▄

vii

Contents

Acknowledgments ▬▬▬▬

WE ARE GRATEFUL to the C.S. Fund, without which this book, and the work that has blossomed from it, would certainly not have been possible. We especially appreciate the continuing enthusiastic encouragement of Herman Warsh, Maryanne Mott, and Marty Teitel, who first suggested that the Oceanic Society produce a comprehensive report on biodiversity in the oceans.

We thank Clifton Curtis for letting us run with this project. His tolerance, as a "small" report on marine biodiversity bloomed into this book, is not forgotten. Also, he and Michael Clark helped to facilitate the publication of this book and the conveyance of its message to national and international forums concerned with the protection of marine ecosystems.

J. Frederick Grassle, John Ogden, Kristina Gjerde, Carleton Ray, George Woodwell, Robert Paine, Richard Miller, Rita Colwell, James Broadus, Michael Weber, and Tundi Agardy shared insights and information that proved invaluable in developing the scientific, political, and economic perspectives of biodiversity in marine ecosystems. Thanks to all these people and to many of our colleagues in the environmental community for their encouragement and suggestions as we put together our report. Thanks to Steven Parcells of the National Audubon Society for providing a forum to test some of the concepts from this book in a practical setting as we joined forces in developing draft guidelines on the funding of foreign development projects impacting marine environments.

Special thanks go to Diane Bennett for her diligent effort in searching for and summarizing the pertinent scientific literature. The many valuable editorial suggestions of Viki Desjardins, Barbara Dean, and Barbara Youngblood have improved and refined the manuscript through its stages of evolution.

Finally, a heartfelt thank you to Sylvia Earle for her moral support and for her belief in the importance of and need for protecting the vast variety of life in the sea. She is one of a very few number of human beings who has extensive, in situ, hands-on, and eyes-on experience with the living world in the ocean—through much of its magnificent depth and breadth. Her scientific and personal impressions of that world give credence to our belief that protected status is appropriate for the interactive network of ecosystems that make up the world's oceans.

Foreword ▬▬▬▬▬▬▬▬▬

COINCIDENT with concerns about the accelerating loss of species and habitats has been a growing appreciation and awareness of the importance of biological diversity to planetary health—and thus to human survival and well-being.

Much has been written recently about the diversity of terrestrial organisms, particularly the exceptionally rich life associated with tropical rain-forest habitats. This is understandable, given the rapid destruction of these areas coupled with the knowledge that, while only about 7 percent of the land is occupied by tropical rain forests, about half the known species occur within these extraordinarily diverse systems.

Relatively little has been said, however, about diversity of life in the sea, although coral reef systems are sometimes favorably compared to rain forests as aquatic examples of species-rich ecosystems. Boyce Thorne-Miller and John Catena's *The Living Ocean* presents for the first time a comprehensive overview of biological diversity in the oceans and associates this with the need to implement programs that will protect marine ecosystems and species. They have taken a view of the planet that would likely appeal to first-time visitors.

Aliens exploring earth would probably give priority to the planet's dominant, most distinctive feature—the ocean. As terrestrial residents, our understandably biased perspective sometimes gets in the way of a true evaluation of global issues. From afar, it is easy to see that land masses occupy only about one-third of the surface. Measured from the standpoint of three-dimensional living space, the ocean part of the biosphere is more than two orders of magnitude larger than the terrestrial part and contains more than 90 percent of life on earth—the bulk of the planet's biomass. Viewed historically, the earth's oceans dominate again, with life in the sea preceding the appearance of terrestrial life by several hundred million years.

One of the great wonders of life is that every individual is different from every other individual. Not just every human being, but every ant, every octopus, and perhaps every leaf on every tree has distinctive quirks and characteristics. There are patterns, of course, and within members of a given species, genetic information flows and maintains a degree of relatedness, of similarity that is perpetuated.

The fact that half of the known species are thought to inhabit the world's rain forests does not seem surprising considering the huge numbers of insects that comprise the bulk of species. Entomologist E. O. Wilson reports finding forty-three ant species belonging to twenty-six genera from a single rain-forest tree in Peru.

Every species is different from every other species, but as wildly different as each ant genus or species may be from every other ant genus or species, their genetic makeup constrains them to be "ants" and, in a broader sense, to share with the 750,000 or so species of insects information that insists they must be insects. The tremendous diversity all reflects variations on the limited themes available and special to the insect.

If basic, broad categories—phyla, classes, and such—are given somewhat greater weight than the splintery ends of diversity recognized as species, then the greatest diversity of life is unquestionably in the sea. Nearly every major division of plant and animal kind has at least some representation in the sea, and many are principally or wholly marine. In contrast, only about half the major divisions of life are represented in terrestrial habitats.

As an example of this broad-division diversity, consider a bleak-appearing rock, measuring about 20 centimeters by 20 centimeters by 10 centimeters and brought to the surface from 200 meters' depth along a steep ocean wall in the Bahamas. At first glance, the rock looks as though it might have been brought from the moon, but upon close inspection, representatives from eleven animal phyla and three divisions of plants are inventoried: several foraminifera (Sarcomastigophora), sponges (Porifera), coral (Coelenterata), a nemotode (Nemotoda), lamp shells (Brachyiopoda), encrusting bryozoans (Bryozoa), a tiny snail (Mollusca), an isopod and several amphipods (Arthropoda), polychaete worms (Annelida), a sipunculid worm (Sipunculida), and a kind of flat crinoid (Echinodermata). Plants include filamentous bluegreens (Cyanophyta), red algae (Rhodophyta), and at least three species of green algae (Chlorophyta).

In a microcosm, this small rock contains much of the history of life on the planet, reflected in the genetic codes of creatures that were more than just a little different from one another. Such large-scale diversity cannot be found in any nonmarine area of comparable size.

To appreciate fully the diversity and abundance of life in the sea, it helps to think small. Every spoonful of ocean water contains life, on the order of 10^2 to 10^6 bacterial cells per cubic centimeter, plus assorted microscopic plants and animals, including larvae of organisms ranging from sponges and corals to starfish and clams and much, much more.

In effect, the entire liquid mantle embracing the planet is a living minestrone, with most of the bits small or microscopic, in addition to a fine assortment of more obvious ocean dwellers—seaweed, shrimp, krill, crabs, fish, dolphins, and a host of little-known categories of plant and animal life.

Most people ignore microbeasts in part, perhaps, because of the size bias that comes from being larger than most creatures. Humans are giants, among the upper 5 percent of species in terms of size, along with whales, horses, and hippopotamuses. Yet, as microbiologists are fond of pointing out, most of the biochemical action that shapes the biological and much of the physical and chemical character of the planet is accomplished by microbes, largely ocean-dwelling microbes. Physiologically, these small creatures, above water and below, are more diverse than plants and animals combined, with capabilities and life-styles that range from being free-living, photosynthetic, and chemosynthetic autotrophs to those that live on and decompose all naturally produced organic materials.

Even on land, the diversity of life is dominated by small, if not microscopic, creatures. Among the nearly million and a half organisms identified, classified, and dignified with a name as distinctly recognized species, insects outnumber all others combined. Some scientists believe that there may be many millions of insect species not yet described, based on the incredible diversity discovered in certain recently explored rain-forest habitats and recognizing that relatively little effort has been concentrated on surveying microfauna, as compared to the attention given to certain favored groups such as mammals and birds.

Imagine, then, the rich diversity of life that awaits human attention in the even less well known realms of the sea. It is possible to gain access anywhere on the surface of the planet and stay for weeks or even years. Repeated visits have been made even to the moon, and instruments have been lofted into space and hurled far beyond our solar system. Yet access to the sea remains extremely limited, even in shallow inshore waters. Only one excursion has been made to the deepest part, which is seven miles down, for twenty minutes, thirty years ago. Even taking into account samples obtained in nets and other remotely deployed instruments, less than 10 percent of the ocean has been sampled, and much of it has not even been more than superficially mapped.

What new discoveries await? As biochemical techniques are refined for determining what constitute plant and animal "species," will the sea be found to contain significantly higher species diversity than presently imagined? Are more "living fossils" cruising the great depths or perched within as-yet-to-be-discovered deep-sea crevices? Will the special kinds of physiology, life-style, reproductive strategies, behavioral patterns, complex interrelationships, and other such factors special to marine organisms be counted as significant diversity issues?

Most important, perhaps, may be the discovery of an enhanced awareness of the need to protect marine biodiversity described by Thorne-Miller and Catena. In a frighteningly short time, the accumulated heritage of all earth history is being modified; some of it is used up, rendered extinct, gone forever. Life is change, and extinction is by no means a new phenomenon, but a single species causing swift and widespread destruction appears to be unique. This type of extinction is surely not in the best interests of that species or of the planet as a whole.

Geologist Don Eicher, in his slim volume *Geologic Time*, suggests a time model with special reference to the significance of life in the sea that gives us a perspective on where we are now. He compresses all of the 4.6 billion years of history into a single year. On that basis, the oldest rocks known date from mid-March. Living things first appeared in the sea in May. Land plants and animals emerged in late November and the richly vegetated swamps that formed the Pennsylvanian coal deposits flourished for four days in early December. Dinosaurs became dominant in mid-December, but disappeared on the twenty-sixth, at about the time that the Rocky Mountains first uplifted. Manlike creatures first appeared sometime during the evening of the thirty-first, and Columbus discovered America about three seconds before midnight.

Viewed this way, not only humankind, but all land creatures are, relatively speaking, newcomers. Broad categories of plants and animals were thriving in the world's oceans long before insects evolved, before birds, mammals, ferns, flowers, trees, and other creatures that most tend to think of as "life on earth."

At least some representatives of most of the divisions of life that have ever occupied space on earth are still represented in the sea—although in some cases by only a few kinds, which are precariously vulnerable to what we do or don't do during the next few decades.

Ammonites, once a large and diverse group of marine mollusks, are now known only from fossils and, by inference, from a few living relatives, the Nautilus species. Unfortunately, Nautilus shells are considered beautiful, and huge numbers are taken for decorative purposes. Merostomates, once a dominant group of marine arachnids and perhaps the ancestors of spiders, are presently known only from four species of the horseshoe crab. Alas, they reproduce in shallow, now largely polluted bays, and the adults are gathered for fertilizer and animal food with little respect for their ancient heritage or precious cargo of distinctively different genetic information. They have persisted miraculously through hundreds of millions of years but may not survive the century.

Most people, even dedicated conservationists, do not view protection of marine species and ecosystems, except perhaps dolphins, seals, whales, and now some coral reefs, with the same sense of urgency accorded to more familiar, seemingly more threatened terrestrial species and systems. The underlying message conveyed by Boyce Thorne-Miller and John Catena is clear in this impressive, important volume. The magnitude of ignorance about the ocean and the diverse life therein is vast; the losses are real, imminent, and permanent. With care, the authors urge, we can, we must, find ways to maintain this irreplaceable living heritage.

Sylvia Earle
Chief Scientist
National Oceanic and
Atmospheric Administration

The Living Ocean

Introduction ▐███████████

IF LIFE ON EARTH has a single outstanding property, it is that it exists in an enormous variety of forms. This was evident to Charles Darwin during the famous voyage that led to his theory of evolution to explain the seemingly unending diversity of life forms that he observed. It is even more evident to those evolutionists who have studied the fossil records and proposed new theories that modify or counter Darwinism. It is sobering to read that more than 99.99 percent of all the species that have ever lived on earth are now extinct.[1] Nevertheless, the array of extant species is enough to amaze anyone who travels the natural areas of this planet, including those who explore the living world within the ocean. Yet, despite this obvious and scientifically tantalizing variety, only a fraction of the species living today are known to science, and many will never be known as living species because they will disappear before they are discovered. This is by now a familiar story for the tropical rain forests, but the story of biological diversity in the oceans is all but untold.

As human beings have populated the lands of the earth, we have pushed out other forms of life. It seemed to some that our impact must stop at the ocean's edge, but that has not proved to be so. By overharvesting the living bounty of the seas and by flushing the

3

wastes and by-products of our societies from the land into the ocean, we have managed to impoverish, if not destroy, living ecosystems there as well. The oceans cover 70 percent of the earth's surface and, when depth is considered, contain on the order of one hundred times more inhabited space than the continents. Unfortunately, so little is known about the variety and distribution of ocean species and about the living processes characteristic of marine ecosystems that it is not yet possible to assess the losses, either qualitatively or quantitatively.

We do know that there is a broader spectrum of different types of life forms in the ocean than on land. This is reflected not in the number of species but in the numbers of higher taxa (families, orders, and phyla), which represent larger genetic differences than occur at the species level. If there are more species on land than in the ocean, as is currently thought to be true, it only means that there are more closely related species on land. How should we weigh the importance of small genetic variations versus very large genetic variations between species?

Besides, the list of marine species that exist has not yet been compiled. Certainly, the number known to science is growing, and it appears that early estimates are far too low. For instance, we are discovering that the deep-ocean floor, originally thought to be biologically poor, supports a diversity of species that may be comparable to that of the tropical rain forests. Almost nothing is known of the numbers and the distribution of species of many microbial organisms that live in the ocean, and recent research has revealed instances where a species described as a single species on the basis of form has turned out to be several species on the basis of molecular genetics. Also, rare species may not be as uncommon in the marine environment as has been assumed; it is more likely that many have simply not been found or identified, because marine ecosystems are so vast, varied, and unexplored compared to terrestrial ecosystems. Species may disappear within the sea without a ripple observed by the human eye.

It is also easier to estimate the damage done by habitat destruction in most terrestrial environments than it is to estimate the damage caused by an alteration in the biological and chemical dynamics of underwater ecosystems. The genetic variety and the vitality of marine ecosystems are suppressed by toxic chemical pollution, eutrophication leading to anaerobic conditions, and overfishing. Impoverished ecosystems can be pushed to the brink of collapse. In such an unhealthy state, relatively small environmental stresses may trigger widespread biological losses, including extinction of species. However, because threats to marine biological diversity are difficult to quantify, they are often simply overlooked.

A major problem in assessing both marine and terrestrial biological diversity is the scarcity of scientists who study the genetic and ecological relationships among living organisms—systematists, taxonomists, and ecologists. Science has recently awarded so much money and prestige to unraveling the secrets of life at the molecular level (where all life is similar) that we have almost forgotten that the differences among earth's life forms are as important as the similarities. However, recent global views of the "living planet" have turned attention to the importance of the variety of biological functions performed by various species, since these functions maintain the geochemical cycles that make earth hospitable to life as we know it. Until we understand these functions more fully and identify the different species, we will not be in a strong position to regulate human activities that are potentially harmful to important elements of the biosphere. In the meantime, it would be prudent for human societies to regulate the activities of their citizens with a precautionary approach—if there is a reasonable chance that an activity will cause serious environmental harm, even in the absence of scientific proof, the activity should be prohibited or modified to eliminate the potential for harm.

An immense and expensive scientific research and training effort lies ahead. Geneticists, taxonomists, and systematists are needed to describe and identify new species in marine ecosystems and to study their distribution, life stages, and gene pools. Ecologists and biogeographers are needed to define the biogeographic boundaries of major ecosystems and to determine the complex biological processes that regulate them. Meanwhile, it is imperative that we restore and preserve their good health and inherent diversity, for they may prove to be the most important regulators of earth's life-support system as terrestrial ecosystems are depleted by human expansion and development. Efforts by science and government combined with environmental awareness and self-regulation by the public will be needed to accomplish the task.

This book presents for the first time an overview of biological diversity in marine environments to help open the way to a new global policy on the relationship of human societies to the living ocean. Biological diversity (or biodiversity) is defined, and the importance of and threats to marine biological diversity are assessed. A review of current scientific knowledge with respect to marine biological diversity is presented, along with commentary on the particular threats to diversity in the various types of marine ecosystem. This is followed by a general discussion of the ways in which government and the public can protect marine biological diversity, with specific examples given of existing national and international policies, laws, and programs. The final chapter includes a view of the future and recommendations for research, policies on economic development and marine conservation, and the role of environmental organizations. It is hoped that this book will be useful to environmentalists and environmental policymakers and managers, as well as to the lay reader interested in the oceans and concerned with the survival of life on earth.

It should be mentioned that the glossary, which includes terms encountered in the scientific review, provides information about some basic biological concepts fundamental to the understanding of biological diversity and ecology. Those with a minimal background in biology may find it beneficial to review the glossary first. The bibliography includes all references used in the preparation of this book, whether or not they are cited in the chapter endnotes.

It is finally important to recognize that this book is an encapsulation of a rapidly growing and evolving body of scientific information, policy and legal frameworks, and management methodology. It is essential that anyone who is working to establish sound marine environmental policy—globally, regionally, or locally—consult the experts who understand the complexities of the ecosystems, the legal systems, and the social systems involved. Environmentalism is an interdisciplinary field where science, economics, law, sociology and politics intermingle. If effective environmental policies—be they marine or terrestrial—are to be crafted, experts from the various disciplines must communicate with each other and with the public, who, in the end, will be the ones to implement the policies.

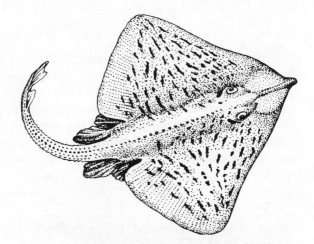

Chapter One ▬▬▬▬▬▬

Is Marine Biological Diversity Threatened?

THE assumption is often made that except for marine mammals (e.g., whales, seals, otters) and sea turtles, species in the oceans are not threatened or endangered; issues of ocean management focus on maintaining a sustained yield of economically valuable species, not on ensuring the survival of species or ecosystems. This assumption is based on the notion that ocean species are dispersed widely. This idea meets no challenge because there is not enough data to determine whether species are endangered or not. However, it is becoming more and more apparent that the diversity of species in the ocean has been underestimated, especially in remote environments such as the deep-ocean floor, and that environmental threats from pollution are increasing rapidly and are spreading much farther than is often assumed by marine policy makers and regulators. Many marine biologists who have observed natural marine communities over a period of time believe they have seen significant declines in populations of some species. But they do not have data to confirm or refute these impressions. More field data is needed if we are to determine the status of marine species. In the meantime,

7

it is important to understand that the absence of knowledge of endangered species or recent species extinctions is far different from
the knowledge of the absence of such occurrences. The fact that
there are so few marine species officially recognized as endangered does not mean that species are not endangered in the
oceans. In other words, we must abandon our assumptions and
gather more facts.

What Is Biological Diversity?

The term *biological diversity* (or biodiversity) has more than one
interpretation, so it is important to understand how the term is
used in a particular context. The purist would use biological diversity to refer only to the number and identification of different species present in a particular ecological system. Because the species
is the evolutionary unit, diversity at this level has special importance. Since it is difficult to count all the species in an entire
ecosystem, several diversity indices have been devised to assess relative diversity in different systems. These indices include both the
species richness (numbers of species per area) and the evenness
(or patchiness) of their distribution. Each mathematical index has
its inherent strengths and weaknesses, but they offer a way to compare species diversity from one ecosystem to another.

In a similar vein, higher levels of classification of animals and
plants may be considered in relation to biological diversity (genus,
family, order, and phylum, in ascending rank). These levels are important when comparing marine with terrestrial biological
diversity.

When comparing species diversity of different ecosystems, it is
important to identify characteristic diversity,[1] which is the species
diversity typical of the unstressed or normally stressed ecosystem.
There are significant differences in characteristic diversity of different types of ecosystems and different geographical locations.
However, it is important not to place greater value automatically
on ecosystems that have higher diversities than others. For example, a salt marsh with a relatively low diversity is not less important
to the living biosphere than a coral reef with a high diversity. What

8

is important is that severely stressed ecosystems tend to have significantly lower species diversity than their unstressed counterparts, and a high value should be placed on maintenance of the greatest diversity characteristic of a particular system.

Recognizing that taxonomic classification is not the only way of categorizing the biological world, scientists now often view biological diversity in terms that include more than just species diversity. For example, the U.S. Congressional Office of Technology Assessment[2] has defined biological diversity to include the following:

- species diversity, which refers to the variety of species in an ecosystem;
- ecological diversity, which refers to the variety of types of biological communities found on earth; and
- genetic diversity, which refers to the genetic variation that occurs among members of the same species.

An additional means of categorizing biological diversity is often discussed—functional diversity. This term refers to the variety of biological processes, or functions, characteristic of a particular ecosystem. This may or may not reflect the species diversity, but it does reflect the biological complexity of an ecosystem and it does identify the nature of processes that may be impacted by human activities.

This fourth category may, in some cases, provide a useful way of assessing biological diversity in the oceans while avoiding the quagmire of cataloging all species (many if not most of which have not yet been identified). By assessing functional diversity, the important biological processes in marine environments can be protected, thereby protecting the organisms performing these functions without having to know how many species there are or what their names are. This may be useful for decision makers, since endangering an ecosystem function may be viewed as more important than endangering a species for which nobody has a particular use. The drawback, however, is that it is likely that some functions of a system will be missed in any inventory. Thus it likely depends upon the ecosystem and the environmental circumstances whether an assessment of species or functions or both would be most useful.

9

The Importance of Marine Biological Diversity

In the face of environmental change, the loss of genetic diversity weakens a population's ability to adapt; the loss of species diversity weakens a community's ability to adapt; the loss of functional diversity weakens an ecosystem's ability to adapt; and the loss of ecological diversity weakens the whole biosphere's ability to adapt. Because biological and physical processes are interactive, losses of biological diversity may also precipitate further environmental change. This progressively destructive routine results in impoverished biological systems, which are susceptible to collapse when faced with further environmental changes.

James E. Lovelock's Gaia hypothesis[3] views the earth, with its biota, rocks, air, and oceans, as a single intercoordinated entity and suggests that the earth's environment has coevolved with its biota. Although much of the theory is controversial and it is better applied on geologic time scales, it provides a useful metaphor for considering interactions between the earth's environment and biota in the short term. There are numerous important feedback mechanisms between the living and nonliving components of the earth.

Among the better-known examples of feedback mechanisms is the potential role of forests—especially tropical rain forests—in regulating global warming. Forests move vast volumes of water from the soil to the atmosphere through a process called evapotranspiration: roots absorb water, which is then moved through the tree to the leaves, where it evaporates through minute openings in the leaf surface. The rate of this process will increase in response to global warming of the atmosphere. This process is predicted to give rise to more clouds, which shade the earth and reduce temperatures, thus moderating the warming effect. When these forests are destroyed, the capacity of the biosphere to maintain favorable conditions is lost. As a result, global warming can proceed at an accelerated pace, and species unable to tolerate or adapt to the temperature change will be lost.

There are numerous examples like this in both terrestrial and oceanic systems. One marine example comes directly from Lovelock and his collaborators. When the theory was introduced, it was proposed that a mechanism to account for the unexplained but measurable movement of sulfur from the sea to the land would be the production of a volatile sulfur compound by marine organisms. Indeed, it has now been discovered that certain species of phytoplankton (floating microscopic algae) in the sea produce just such a compound—dimethyl sulfide (DMS)—in large quantities. Furthermore, it has been suggested that production of this compound will increase as average global temperatures rise due to the greenhouse effect; the substance, in turn, forms an aerosol with droplets that act as nuclei for cloud formation, and the increased cloud cover will have a cooling effect.[4]

There is further evidence of the interaction between ocean biota and global climate change. The geological record in ocean sediments and polar ice sheets suggests that variations in both the carbon dioxide and the climate are linked in some way to changes in the amount of carbon being taken out of the atmosphere and laid down, as organic matter or shells, in marine sediments. It is not known just how climatic change and change in the ocean carbon cycle are related, but there has been scientific speculation that increased productivity of the oceans has caused global cooling by reducing atmospheric carbon dioxide.[5]

In light of biological feedback mechanisms and the importance of marine biological communities to the geochemical cycles of the planet, efforts to save biological diversity at some level can be justified on stronger grounds than the one often proposed, e.g., the potential sources of food and medicines for humans. The very existence of life as we know it, and certainly the human species, depends on the ability of the biosphere to tune the chemical and geochemical cycles of the earth.

In addition to the ecological importance of maintaining the diversity of biological functions in ecosystems—both terrestrial and oceanic—there are more specifically human-centered arguments for maintaining biological diversity. Among the potential uses of a wide variety of species are the following: source of food; genetic stock for agricultural, aquacultural, and sylvicultural breeding; source of genetic material for biotechnology applications; source of important drugs and other medical uses; and source of materials used in food processing and other industrial applications.

11

The importance of food from the sea varies considerably from nation to nation. In the United States, for instance, seafood tends to be highly priced and is generally not a staple food item, but it provides a livelihood for many fishermen serving both local and export markets. Coastal fisheries are seasonal and often local, and are highly susceptible to human impacts, especially pollution. Offshore fisheries supply national and international markets and suffer more and more from overfishing. In contrast, in many of the coastal and insular developing countries, seafood is a staple and constitutes a large portion of the economy. These fisheries are susceptible to overfishing, bad management practices, and pollution from development. Six of the top 11 fish-harvesting nations are developing countries (China, Chile, Peru, India, Republic of Korea, and Indonesia). The most heavily fished areas in the world oceans are the northeast Atlantic and the northwest Pacific, which are fished by technologically advanced fishing fleets from Europe and Japan, respectively. In fact, the top five fish-harvesting countries of the developed world frequent these waters (Japan, the Soviet Union, the United States, Norway, and Denmark).

The World Resources Institute estimates that about 9,000 species of fish currently are exploited, with only 22 of these harvested in global-scale quantities (exceeding 100,000 metric tons annually). Five types of fish (herring, cod, jack, redfish, and mackerel) account for about half of the annual world catch. There was a tremendous growth in fisheries in the sixties and seventies, but that growth has slowed down or reversed due to overfishing.[6]

Coral reefs are ecosystems of the highest diversity that support fisheries, and fishing pressures are intense in most reef areas of the world. Unlike the fisheries based on populations of schooling fish, such as sardines and tuna, the reef fisheries are based on numerous nonschooling species, each of which exists in relatively small populations. Although these systems support a high diversity, the productivity of any one species is not high enough to support intensive fishing. This is especially a problem where an export fishery is focused on only a few species, such as jack, grouper, snapper, and lobster, which are rapidly overfished. However, this is not the only problem coral reefs are facing. Fish and invertebrate communities in nearly all reef areas are being overexploited because local fish-

12

eries are feeding expanding populations and because of extensive collecting for aquarium and seashell trade.

During the fifties and sixties, there was a great deal of public speculation about how the oceans would solve future world food problems. It was predicted that new food sources would be exploited and that large-scale sea farming would become popular. This prophecy has not turned out to be true, and such speculation is no longer popular. Most fisheries have reached or surpassed their sustainable yield and many are severely depressed. A few species formerly treated as "trash fish" have been marketed, but there has been little other development of new food sources. Sea plants have not proved to be a major food worldwide, as once hypothesized, but they continue to be harvested and marketed in cultures that have historically used them (e.g., Japan). The market for sea plants in other countries has increased, but primarily as a specialty item in "health food" or "gourmet food" outlets.

It is not uncommon for marine organisms to produce toxins; these are functional in repelling predators or in giving the species a competitive advantage by releasing a substance that retards the growth of another species. As a direct consequence of this "chemical warfare," there is a wide variety of biologically produced substances from the sea that have medicinal value for humans. Derivatives from marine flora and fauna include numerous drugs with various properties: antibiotics (both antimicrobial and antiviral), tumor inhibitors, coagulants and anticoagulants, and substances used in treating heart or nerve ailments.

Examples of marine organisms with potential medicinal substances include several algae that have antibiotic and anticarcinogenic properties. Corals, sea anemones, sponges, and mollusks all include species exhibiting antibiotic activity. The porcupine fish and the puffer fish have yielded symptomatic treatments for terminal cancer, and shark's liver contains substances that enhance resistance to cancer. Sea cucumbers, sea snakes, menhaden, and stingrays produce materials useful in the treatment of an array of cardiovascular ailments, and extracts from seaweeds and octopus treat hypertension. Antiviral substances isolated from seaweeds are active against viruses causing cold sores, eye infections, and venereal disease, and a sponge has yielded a substance effective against

13

viral encephalitis. Thus, it may be that the sea will lead the way to the successful treatment of viruses, including the common cold.[7]

Seaweeds have proven particularly useful to humans even beyond their importance as sources of drugs. Their nutritive properties, especially their vitamin and mineral content, make them useful as food or as food additives. Other seaweed derivatives are used in food processing—alginates, agars, and carageenans being the most common of these—and in the production of cosmetics, shampoos, and detergents. In these applications, seaweed extracts are most commonly used as thickening agents and emulsifiers. They are also found in pharmaceutical products as nonactive ingredients or digestive aids. In addition, seaweed tissue may provide a source of fibers, plastics, waxes, lubricants, and even paper.

With the advent of genetic engineering there is an opportunity to enhance production of marine metabolites useful in medicine. Often the need for important genes for biotechnology is used as a justification for preserving genetic and species diversity. For example, one scientist has written that biotechnology offers a "strong foundation for exploitation of biologically active compounds already known to occur in the sea and for further exploration into the recesses of the world oceans for compounds and food sources as yet undiscovered."[8] However, the dangers of genetic manipulation should be recognized, and biotechnology may prove to be as much a threat to natural species and genetic diversity as it is a justification for maintaining that diversity. The release of individuals with artificially composed genetic makeup into wild populations of the same species could upset the natural distribution of that species as well as the competitive interactions with other species, destabilizing natural biological communities.

The Threats to Marine Biological Diversity

Biological diversity has become widely recognized as a critical conservation issue only in the past two decades. The rapid destruction of the tropical rain forests, which are the ecosystems with the highest known species diversity on earth, has awakened people to the importance and fragility of biological diversity. The high rate of species extinctions in these environments is jolting, but it is important to recognize the significance of biological diversity in all ecosystems. As our numbers continue to expand, the human impact on biological diversity will overtake one after another of the earth's ecosystems. In terrestrial ecosystems and in fringe marine ecosystems (such as wetlands), the most common problem is habitat destruction. In most situations the result is irreversible. Now we are beginning to destroy marine ecosystems through other types of activities, such as disposal and runoff of poisonous waste. In less than two centuries, by significantly reducing the earth's gene pool, we have unraveled eons of evolution and irrevocably redirected its course.

Certainly, there have been periods in the earth's history when mass extinctions have occurred. The extinction of the dinosaurs' was caused by some physical event, either climatic or cosmic. There have also been less dramatic extinctions, as when natural competition between species reached an extreme conclusion. Only .01 percent of the species that have lived on earth have survived to the present, and it was largely chance that determined which species survived and which died out.

However, nothing has ever equaled the magnitude and speed with which the human species is altering the physical and chemical world and demolishing the biosphere. In fact, there is wide agreement that it is the *rate of change* we are inflicting, even more than the changes themselves, that will lead to biological devastation. Life on earth has continually been in flux as slow physical and chemical changes have occurred on earth, but life needs time to adapt—time for migration and genetic adaptation within existing species and time for the proliferation of new genetic material and new species that may be able to survive the new environment. However, the swift attack by humans does not allow time for evolution, only for biological wreckage.

The biological diversity of both terrestrial and marine environments is being threatened, but there are major differences in the nature of the threats, and in the effects of those threats, on the two major living realms of the earth: the land and the ocean. Because of these differences, the assessment and protection of the two environments cannot be handled in the same way.

The ocean environment is gentle compared to that of the land's.

15

Oscillations from season to season and from year to year are muted, especially in the deep ocean far from the sea's edge. The multitude of life forms have sorted themselves into zones of preference in accordance with latitude and depth, and within each zone the plants and animals have had stable conditions for extended time. This stability has allowed the proliferation of species, but many if not most of these species are not adapted to deal with sharp changes in their environment on either seasonal or long-term scales. Now they are threatened because they may not be adaptable enough to deal with rapid environmental changes occurring globally. The most adaptable marine species are those in intertidal and estuarine environments, where natural fluctuations in the physical environment are most pronounced, but they represent only a tiny fraction of the species in the ocean.

Ironically, it is a terrestrial species, *Homo sapiens*, that now threatens life in the oceans. Human excesses—the overwhelming growth of populations and the unchecked wastefulness of "advanced" societies—are causing environmental disturbances that cascade from land to sea at a perilous rate. Life in the sea is vulnerable to the environmental consequences of our lives.

Habitat Destruction

In the oceans, there is critical habitat destruction in coastal areas where humans build on, cut through, dredge out, or bury entire marine communities. Port and harbor development, industrial facilities, tourist facilities, mariculture development, and expansion of urban areas are causing the irretrievable loss of critical coastal habitats. The United States, for example, had lost roughly 50 percent of its original coastal wetlands by the mid-1970s.[9] Similarly, mangrove ecosystems in tropical countries face tremendous development pressures because of their value as firewood and domestic fuel, and from conversion to brackish-water shrimp ponds and reclamation for housing and industry.[10] Moreover, coral reef ecosystems are facing a variety of pressures. In Southeast Asia, for example, coral reefs are being destroyed by blast fishing. This technique uses a variety of commercial, military, and homemade explosive devices and is a significant problem in Malaysia, Thailand,

Indonesia, and the Philippines.[11] Vast areas of the oceans far removed from the effects of expanding human populations remain physically undisturbed. However, the threat of habitat destruction will grow significantly if mining the seabed for minerals becomes a widespread activity. A similar threat could come from seabed burial of wastes, large-scale dredge spoil disposal at sea, and the futuristic but not impossible prospect of constructing floating islands for industrial and residential development.[12]

Water Pollution

The question now often raised is whether biological diversity is, in fact, threatened in those marine environments where habitat destruction does not play a major role. Because species of ocean plants and animals are not well known, especially from such remote environments as the deep-sea floor, it is difficult to know whether and how many species are facing extinction. The conditions of even the more familiar coastal species are difficult to assess, simply because their populations are under water, where they cannot be readily surveyed. There are, however, anthropogenic threats to marine biological diversity even in deep-water ecosystems far removed from the footsteps of humankind. There are clear indications that pollution has spread even to these remote environments.

It is the basic nature of the oceans—the water—that makes pollution such a threat in marine environments. The chemical outfall of advanced civilization is soluble and reactive in water. Consequently, organisms that live and feed in water are susceptible to the effects of these chemicals. In water, the chemistry of pollution reacts readily with the chemistry of life.

Pollution from land is threefold: dissolved nutrients, dissolved toxics, and suspended particles (toxic and nontoxic). These wash into the oceans with runoff from agriculture, urban/industrial activities, deforestation, and construction, and through direct sewage outflow. Atmospheric pollutants are also proving to be a major source of coastal pollution, both nutrients (e.g., nitrogen) and toxins. For example, it has been estimated that 25 percent of the nitrogen-containing pollutants entering Chesapeake Bay are from

17

airborne sources—deposited directly or washed in with rain.[13] Estimates of the proportion of toxic pollution from atmospheric deposition run as high as 40 percent in Europe's coastal waters and greater in the open oceans, but the data is sparse and the estimates rely to a great extent upon inference.[14]

The fate of pollutants in marine environments is varied and complex. Some portion remains circulating in the water column, and may be persistent or may break down into other products over time. Any toxin dissolved in the water is available to the food chain, where it does its damage. The sediments underlying coastal waters attract and accumulate contaminants, often to extremely high levels. These high concentrations are a direct threat to animals that dwell in and on top of the sediments—and also to life in the overlying water as the contaminants slowly leach from the sediments.

The biological consequences of pollution are predictable and are repeated over and over again in coastal ecosystems around the world. In a process called eutrophication, nutrients (especially nitrogen and phosphorus) from sewage, agriculture, and atmospheric pollution cause excessive microscopic plant growth, which is followed by a decline in dissolved oxygen as it is used up during microbial decomposition of dead plant material. The low oxygen leads to the demise of intolerant species. Increases in water turbidity due to increased loads of suspended sediments or excessive blooms of microalgae diminish the light penetration into coastal waters, thereby threatening many light-loving species. Eutrophication invariably results in reduced species diversity as a few very successful species outcompete or drive out other species. It is interesting that the blooms of microscopic plants are not only low in species diversity; they are also frequently dominated by toxic species (sometimes known as red tides). They not only foul the water; they also foul the food chain.

Toxic pollution may kill marine organisms outright, but sublethal effects are more prevalent—disease, deformities, accumulation of high levels of toxins in living tissue, and a weakening of the genetic pool (by killing off the individuals whose genetic composition does not enable them to adapt to the stressful conditions). Through a process called bioaccumulation, organisms concentrate noxious materials over time. When low concentrations of toxins are supplied continually, organisms eventually may suffer from the high concentrations they have accumulated. Often the effects are sublethal. Chronic ailments include deformities, growths, lesions, and diseases that do not kill the organism immediately. In another process, called biomagnification, persistent toxins are passed along the food chain and accumulate in progressively higher concentrations; so animals that feed high on the food chain risk very high levels of tissue contamination. It is a general rule that as pollution increases, the physiological and genetic health of marine populations declines, and eventually the number of species in the polluted ecosystem declines.

Because of the fluid and turbulent nature of the oceans, pollutants entering marine environments are widely dispersed from their point of origin. In addition to the physical dispersal, there is a biological dispersal through food webs. Some scientists and environmental managers refer to the dissipation of toxic pollutants—from the source broadly and deeply through the water and into the sediments—as dilution and assimilation, and they argue that the effects of pollution are in this way reduced or avoided. However, it is more accurate to call it dispersion and to acknowledge that in this way the effects are broadcast, even to the mid-ocean and deep-sea floor. Living organisms play a major role in the ocean's "assimilation" of toxins, and they are often weakened or lost in the process.

Marine Debris

Another type of pollution that plagues life in the oceans is marine debris, including plastics, fish nets, and many other forms of trash and garbage. Seagoing vessels and seashore activities are sources of this enormous problem, which exists even in the most pristine environments.[15] The lives of fish, sea mammals, seabirds, and sea turtles are threatened by debris that may entangle and drown or trap them or, if ingested, poison or choke them. It is not known to what extent minute particles of this debris threaten smaller forms of sea life, but it is reasonable to expect the threat to be real.

Overharvesting

Another important threat to marine ecosystems and their diversity is overharvesting. Our fishing and hunting methods have caused quantitative, genetic, and social disruption of populations of a multitude of marine fish, shellfish, turtle and mammal species. The effects of overfishing of selected marketable species are widely recognized, but the concomitant depletion of species incidentally caught and discarded also has serious consequences to complex marine food webs. Mismanagement or lack of management has depleted populations of some of the major fisheries of the world. For example, the depletion of the anchovy fisheries off the coast of Peru during the seventies resulted in a loss of 7.5 billion metric tons in one year—or approximately 80 percent of its potential had good management practices been in force.[16]

Overfishing and a change in the southern Pacific Ocean's weather and ocean current patterns combined to cause a devastating impact on the Peruvian anchovy fishery. During seven of the eight years from 1964 through 1971, the anchovy harvest exceeded 9.5 million tons, the level identified by a team of biologists from the United Nations Food and Agricultural Organization (FAO) as the maximum sustainable yield of the fishery. The harvest subsequently plummeted to about 4 million tons, and then to just 1.5 million tons in the early eighties.[17] It remains questionable whether or not the fishery will ever return to its level of two decades ago.

Species diversity may be affected beyond the loss of harvested species. If any of those species has a pivotal role in its ecosystem, its absence or reduction will dramatically affect the community structure, often with the result of a reduction in species diversity. Perhaps most pathetic, from our anthropomorphic point of view, has been the disruption of family and social structure among marine mammal species, many of which have been hunted or killed to near extinction. Damage to the food chain may also threaten these creatures, who either rely upon great quantities of plankton or feed high on the food chain.

Global Climate Change

In addition to direct stresses placed on marine environments by human activities, there are indirect effects from the global environmental changes being wrought by industrial pollution. Most notorious of these is global warming due to the "greenhouse effect," in which accumulations of carbon dioxide and other gases (such as methane) trap solar infrared light (heat) in the atmosphere. There is a great deal of scientific argument as to whether the warming is upon us or is still to come, but all models predict some warming (even if all anthropogenic sources of carbon dioxide were to be halted today). The degree of temperature increase is as yet unknown and will depend upon various factors, such as the role of the oceans in removing carbon dioxide from the atmosphere and the possible increase in cloud cover resulting from biological responses to warming and increased carbon dioxide. Atmospheric warming is expected to be greatest at higher latitudes in the Northern Hemisphere.

The rise of ocean temperature will not be as fast or as great as atmospheric temperature because of the high heat capacity of the ocean. However, only a degree or two can have a dramatic effect on biological communities. There will also be other effects, such as a predicted sea-level rise due to thermal expansion of the oceans and to melting of the Arctic tundra and ice cap. Among the possible effects on sea life are: (a) the significant loss of coral reefs, salt marshes, and mangrove swamps unable to keep up with a rapid rise in sea level; (b) the loss of species whose temperature tolerance range is exceeded, which is perhaps an even greater threat to corals than is sea-level rise; and (c) effects from the tundra runoff, which could include runoff of nutrients and suspended sediments. Sea-level rise will also result in saltwater intrusion that wreaks havoc with freshwater ecosystems, including rivers, freshwater marshes, and coastal lowland farm acreage.

Recognition of the Threats to Marine Biological Diversity

The variety of human assaults from land and sea, as described above, has led to impoverishment or a weakened state for many marine ecosystems. Numerous species, while not extinct, exist in greatly reduced and scattered populations. Their gene pools have been severely reduced and, therefore, so has their ability to adjust to rapid environmental changes that may occur in the near future. They are also physiologically weakened by chronic toxicity caused by numerous pollutants. In short, many marine species are surviving in a highly susceptible state and are more vulnerable to any new environmental stresses. It is therefore difficult to estimate how endangered a species may be. Extinction of some of these impoverished species might occur as a result of rapid population collapses rather than as a slow dwindling of numbers.

Unfortunately, the threat to marine species has not been recognized in official tallies of endangered and threatened species. Obviously, because of the difficulties in identifying and monitoring marine species, the species-by-species approach to protecting diversity will not work for the oceans. Of tabulated numbers of endangered and threatened species worldwide, 16 out of 6,691 are marine (not including shorebirds) and of those 14 are mammals and turtles.[18] Such an evaluation process could easily and erroneously lead one to the conclusion that biological diversity in the oceans is not threatened.

While the threat to marine biological diversity per se has not been recognized, the generally deteriorating state of coastal waters due to pollution and coastal development is now acknowledged worldwide. What to do about it, if anything, is a subject of debate in national and international forums, from parliaments to global conventions. Since the causes of this degradation are also the major causes of loss of marine biological diversity, effective measures to stop the one will help protect the other. The tone for the future has been aptly set in the 1987 report by the World Commission on Environment and Development, entitled *Our Common Future:*

> Looking to the next Century, the [World Commission on Environment and Development] is convinced that sustainable development, if not survival itself, depends on significant advances in the management of the oceans. Considerable changes will be required in our institutions and policies and more resources will have to be committed to oceans management.[19]

Chapter Two ████████

How Science Defines Biological Diversity

I N studying and describing biological diversity, ecologists have defined a number of terms and concepts related to the complexity and stability of ecosystems.[1] The following terms are used to describe community complexity:

- *Species richness* refers to the number of species in a system.

- *Species evenness* is a measurement of how evenly species are distributed among themselves and how evenly the populations of each species are distributed.

- *Interspecific interactions* are direct interactions between species, such as competition and predation. An estimate of the average number of other species with which one species interacts directly is consistently around 3–5, with values in relatively constant environments higher than in fluctuating ones.[2]

To have a numerical means of comparing diversity in biological communities, species diversity may be represented by a mathematical *diversity index*, which expresses the relationship between number of species and their relative abundance or evenness.

To describe community stability, the following terms are used:

- *Stability* is environmental constancy, or persistence of a community in time (little fluctuation in population). A system is deemed *stable* if and only if the biological variables (species diversity, food webs, productivity, etc.) in the system all return to the initial equilibrium following a disturbance.

- *Resilience* refers to how fast the variables return toward equilibrium following perturbation. While useful generalizations can be made about the relative resilience of different systems, it is important to realize that the strength of the disturbance felt by the system can be difficult to predict. For example, an ecosystem might return rapidly to equilibrium after a major short-term pollution event but very slowly after a period of long-term low-level pollution. Resilience in marine systems can be complicated by the availability or nonavailability of pelagic larvae for recolonization.

- *Resistance* refers to the degree to which the ecosystem is changed following a disturbance.

- *Variability* is the natural change in population densities over time.

The characteristic diversity of an ecosystem refers to the species diversity attained when the ecosystem is in a relatively unstressed state (that is, when it is exposed to disturbances that are normal for that type of ecosystem). As a system loses its characteristic diversity, which often happens when alien stresses are applied, the stability of the ecosystem is jeopardized. Sequential stresses thus become progressively more damaging.

Keystone Species

Especially in benthic communities, certain species have functional roles that are more important than their abundance or biomass suggests. These are known as keystone species, and fluctuations in their populations can have significant impacts upon the entire community in which they live. Other species may come and go without fundamentally changing the ecosystem processes. It is critical to identify the keystone species in an ecosystem so that they can be used as indicators of the health of that system.

An example of a marine keystone species is the starfish *Acanthaster*, or crown of thorns, on Pacific coral reefs, especially the Great Barrier Reef. Due to its efficiency as a predator and its ability to disperse its population over a large area, increases in the starfish's population can cause widespread devastation of a coral reef system.

Another well-known keystone species is the sea otter in kelp ecosystems. Sea otters feed heavily on sea urchins, the spine-covered invertebrates that graze on seaweeds (in this case, on the kelp). The kelp provides the physical structure that supports the biological diversity characteristic of that type of ecosystem. Actually all three—the sea otter, the sea urchin, and the dominant kelp—are keystone species in that the disappearance of one will change the whole community. If, for instance, the sea otter is removed, the sea urchins proliferate and eventually eat away much if not most of the kelp (depending in part on the presence of other urchin predators), and many species that depend upon the physical environment provided by the kelp then disappear as well. This happened in many locations when the sea otter was hunted for its fur—from the dense kelp beds along Alaska's rocky shores to the majestic kelp forests off the California coast.

Genetics

Because the dynamically circulating waters of neighboring ecosystems are usually exchanged to some extent, many marine species have developed a great dispersal capacity. Thus, it would be natural to expect great genetic interchange among populations of marine species.[3] In fact, modern biochemical techniques, which can identify gene structure within individual chromosomes, have revealed that genetic differences between separate populations of many species are significant. The separation is usually spatial (or geographical), but for some species there can be a temporal separation of genetically distinct populations. For example, one species of phytoplankton may have two distinct populations, each blooming (reaching a seasonal population peak) at a different time of year in the same ecosystem.[4]

Factors restricting genetic flow between populations may include: environmental barriers to larval survival—i.e., larvae from one ecosystem may not find the conditions necessary to support them in another ecosystem so that the genetic variations they carry do not survive in the new location; and environmental barriers to reproduction at the new site even when dispersal has effectively introduced immigrants to a new population. Such exclusions, however, are apparently not so complete that they result in speciation (the evolution of new separate species). There must be enough successful crossbreeding to maintain the integrity of the species, but different environmental conditions favor different dominant gene combinations within the respective populations of the species.

This phenomenon suggests that, at least for some marine plants and animals, species may not be the significant unit of biological diversity. For these organisms, it is the diversity of populations (also known as genetic diversity) that has to be protected. This illustrates how the endangered species approach, or even the habitat protection approach, may fail to give adequate protection to important members of marine ecosystems. If a discrete population is depleted, critical genetic information may be lost; this might not "endanger" the species but could result in its exclusion from certain habitats it had once occupied.

When trying to relate the biological functions that characterize an ecosystem to species diversity, it is important to understand that there is not always a direct correspondence between the two. Species may be specialists, performing a narrow range of functions, or they may be generalists, performing a wide range of functions. Specialist and generalist species can be characterized genetically as well as functionally. The categories relate to the amount of genetic variation found in a typical healthy population. Specialist species have narrow environmental requirements and are found only in highly specific ecological niches. This trait tends to be associated with high genetic variability among the individuals of a population. Generalist species, on the other hand, have broad environmental requirements and are found in broad niches composing a variety of functions, or life-styles. A generalist species, therefore, may be

26

found in widely varying environments or in an environment that has large fluctuations. These species tend to have low genetic variability among the individuals of a population.

In a study of marine teleosts (fishes with a bony rather than a cartilaginous skeleton), a negative correlation between genetic variability and environmental range was found.[5] High genetic variability was found in specialist species, which are characteristic of the following habitats: tropical, temperate pelagic, and intertidal/sublittoral. Low levels of genetic variation were found in generalist species characteristic of habitats such as temperate demersal and polar. Thus, the characteristic genetic diversity must be taken into consideration when comparing species in different ecosystems and when determining whether a species in a given ecosystem is genetically impoverished.

In another study, however, where marine phytoplankton populations were genetically assessed, no relationship was found between the amount of genetic variability in the populations and the apparent variability or predictability of their environments.[6] But it was noted that the amount of environmental variability or predictability perceived or conceived by humans may not reflect the amount experienced by phytoplankton cells, so generalizations concerning genetic variability in natural populations may not be correct.

Understanding the genetic structure of a population is particularly important for species used in pollution monitoring. For example, a study found unexpected genetic variation in what was thought to be a single species of *Capitella,* a genus of marine worm that is often used as an indicator of pollution. Genetic analysis revealed that what had been recognized as a single species is actually comprised of at least six species.[7] It was concluded that the differences in the life histories of the different species may provide even more sensitive indications of patterns of disturbance such as pollution. Furthermore, it may suggest that there is a significant amount of hidden species diversity in the ocean.

In a study of another indicator species, the fourhorn sculpin (a fish), it was noted that the effects of pollutants on natural populations cannot be accurately assessed without knowledge of the genetic structure of the species. If there is spatial and/or temporal variability in genetic structure, the responses of different populations to pollutants may be quite different.[8] Examples of isolated populations of species adapted to highly toxic, polluted environments have been found. Such a population appears normal and indistinguishable from similar populations in unpolluted environments, but in fact it has a different genetic structure.

Comparison of Biological Diversity on Land and in the Ocean

Of the species known to science, about 80 percent are terrestrial, but there are more orders and phyla in the sea.[9] In fact, all phyla of animals are found in the sea, a majority of these in benthic environments, and one third of the phyla are exclusively marine.[10] If plants and protista are also considered, at least 80 percent of all phyla include marine species. In addition, the relative abundance of marine species may be considerably greater, since most marine species are unknown. For example, bacteria and viruses and other primitive species are just now being recognized as major components of the marine biota. Identification of these species is difficult and is being done at the molecular level by analysis of protein sequences in the DNA molecules of the chromosomes.

In order to count species, certain kinds of information are needed, which may be difficult to acquire with the present knowledge of life in the oceans. Even the most basic information—e.g., how different two groups of organisms have to be before they are called different species—has recently been called into question by genetic studies on certain marine species.

It is also difficult to compare the number of species in different orders or phyla, because some taxa have been studied in far more detail than others. To date, more mollusk species (including most "seashells") have been identified than any other group of marine animals, but this could reflect the fact that these species are more accessible and better studied because many of them live in intertidal or shallow water, and those with durable shells can be retrieved in good shape from the deep-water habitats they may occupy.

Some scientists suggest that simply counting species is not a valid way of assessing an ecosystem, marine or terrestrial, because it can be a misleading measure of diversity and does not represent the whole genetic picture. Carleton Ray, consistent with the functional analysis approach, recommends examining "life-styles" (what they do instead of who they are).[11] There are several life-styles of major importance in the ocean that are totally absent from land. For example, filter feeding—practiced by barnacles, baleen whales, and others—relies upon a fluid matrix and is therefore not a life-style found on land.

Also indicative of greater diversity of functions among marine biota is the fact that marine food webs tend to be more complex, with more trophic levels, than terrestrial ones.[12] This suggests that there are feeding behaviors or regimes that exist only in the sea. This may seem inconsistent with the claim that there are many more species on land than in the oceans. However, apparently there are fewer marine species functioning at the same trophic levels than there are on land, and/or there are more terrestrial species that have nearly the same function. The environment may not be partitioned as finely among species in marine environments, so that fewer species perform more functions.

One factor that possibly contributes to a higher species diversity on land than in the oceans is the large vegetation forming a structured environment. This is absent in most marine systems, except wetlands, kelp beds, and reefs. The most elaborate of these structures can be found in coral reefs, where animals (corals) build an organic structural matrix of calcium carbonate. This matrix leads to increased diversity due to increased spatial complexity and increased biological interactions, so some of the highest-recorded diversity in the sea exists on coral reefs. However, systems dependent on a destructible matrix are less resilient than systems lacking such a matrix, because disturbances may reduce or eliminate the structure upon which the diversity is based.[13]

Another common claim is that endemism is uncommon in the oceans as compared to land environments, and this may in part account for fewer marine species. Marine species, so the story goes, are only rarely limited to small defined areas—areas usually associated with isolated physical or geochemical structures such as reefs, trenches, seamounts, or hydrothermal vents. Even in these environments a surprising majority of species seem not to be endemic. The oceans provide a circulating medium ideal for the dispersal of spores and larvae, so it is not unreasonable to expect species to be widespread. Acceptance of this rule generally leads to the conclusion that extinction of species is less likely in the oceans.

This argument has two potential sources of error. First, endemism, or rareness, may not be as uncommon as thought. Ocean species have been so poorly studied that many rare species may not have been discovered. Certainly, recent studies from remote areas such as the deep sea are revealing a heretofore unimagined richness of species and a more patchy distribution than expected. Second, species may not be the critical unit for genetic extinction in the oceans. Defined populations with limited distribution may contain the kinds of genetic variation that species have in environments like the tropical rain forests. Thus, if populations are extinguished, important gene pools may be lost.

Endemism aside, it is certainly true that many or most marine species are characterized by widespread distribution as a result of their modes of reproduction and the ability of the fluid ocean medium to broadcast spores and larvae. This has implications for organisms, such as many invertebrates, that are benthic during the adult stages of their lives. The restocking of these populations by larval influx and settlement is an important mechanism for maintaining marine benthic communities. Benthic species may be short-lived, or physical disturbances may cause patches of benthic habitat to become available for recolonization at frequent intervals. The influx of larvae carried on ocean currents from other areas provides the stock to replenish such communities with the same species or with other species that find the environment favorable.

In fact, the reproductive cycles of most marine animals, benthic or pelagic, are related to the fluid circulation patterns of the ocean. Successful reproduction in most species depends on a combination of physical dispersal and relatively predictable food cycles. Due to physical dispersal, early life stages are spatially separated from the adult state—unlike most terrestrial populations (especially vertebrates), whose reproductive strategies place adult and juvenile life stages in close proximity to one another. Physical dispersion from one ecosystem to another can blur boundaries between those ecosystems, and larvae from one may seed or provide food for a neighboring system. On land, in contrast, crossing from one ecosystem to another requires active migration and not passive dispersal.

Differences in Physical Properties of the Environments

Several physical differences between marine and terrestrial environments play a role in determining relative species diversity in the two realms. One of the major differences influencing the biota is the fluid nature of the oceanic environment, which, in addition to enhancing cross-fertilization and dispersal of marine species, dissolves and circulates nutrients. Patterns of nutrient distribution in the ocean tend to enhance biological production in some areas more than others. The circulation of ocean waters also often facilitates the broad distribution of toxic pollution, threatening biological communities in its path.

The air/rock-dominated systems and water-dominated systems also differ with respect to the magnitude of natural environmental fluctuations. The strong seasonal and interannual fluctuations in the terrestrial climate contrast with the moderate fluctuations in the marine environment. The moderation of seasonal differences in marine environments is due to the permanent wetness of the environment and the great heat capacity of the ocean—it takes a large input of heat to make the water temperature rise only slightly, and vice versa. These differences in temporal variability of the environments may result in major biological differences.

The large environmental variability over both short and long term suggests that terrestrial organisms have developed physical or physiological mechanisms to cope with short-term variability, and this in turn may serve to minimize the effects of long-term variations.[14] In contrast, environmental variability in marine systems is small, and extends over very long time scales as a result of the long periods of exchange between deep and near-surface waters (100–2,000 years) as well as the large thermal capacity. Marine systems require less robust internal processes to respond to low-magnitude short-term variations; therefore, marine systems would have different responses, or less ability to respond, to long-term variations. In short, marine ecosystems may be more vulnerable to large-scale environmental changes, such as pollution and climate change, because they do not have the internal adaptability inherent in terrestrial systems accustomed to environmental oscillations.

The moderation of temperature extremes in ocean waters has resulted in the dominance of poikilothermic physiologies (animals that do not regulate their own body temperature but take on the temperature of the environment). There is little need for individual pelagic organisms to create their own internal or immediate environment, and therefore warm-blooded animals have not evolved in the sea. The only ones found there, the marine mammals, actually evolved from terrestrial mammals.

Terrestrial environments have more pronounced physical boundaries between ecosystems. The fluid nature of the oceans does not prevent the formation of distinct ecosystems, but the boundaries between ecosystems are softer than on land. Except for the very sharp boundary between land and sea, the boundaries setting apart marine ecosystems are usually defined by currents or strong gradients in physical or chemical properties, such as light, temperature, and salinity. Although ocean waters are contiguous, currents form subtle boundaries that separate water masses with differing environmental conditions. Currents can act as physical barriers to the small organisms that drift with moving waters, or to the larger, motile organisms that find environmental conditions on one side of the boundary more favorable than on the other side.

Great depth is another important property of the ocean distinguishing it from the terrestrial environment. This expanded third dimension provides a structure that can support a variety of life forms that do not exist on land. Species are distributed in vertical zones in response to environmental gradients (light, temperature, oxygen) and food supply.

Marine plants obviously have to remain within sunlit waters (the photic zone). Some are most commonly found in surface waters of high light intensity, while others proliferate at depths of lower light intensity. Characteristically, biological communities are defined by species vertically clustered in distinct assemblages rather than species independently distributed along a vertical gradient. For example, in the central Pacific Ocean, phytoplankton species are assembled into two communities: one in the top 80 meters or so, the other below that depth to where there is not enough light for photosynthesis. Similarly, on rocky coasts, there are several distinct communities distributed with depth, and at low tide distinct color bands can be distinguished, each color reflecting the dominant species in a community.

Animal species also separate themselves into different vertical zones according to their feeding habits: herbivores in the upper strata; carnivores at various levels, depending on their particular prey; and detritivores in deep waters and on the bottom. Competition between species and pressures of predation may also influence the vertical distribution of species. Thus, depth provides a greater ecological diversity as well as more complicated food webs, and these in turn mean increases in functional and species diversity. The structure provided by the seawater itself makes up to some extent for the lack of biological structure (e.g., trees) that supports diverse communities on land.

A special case of vertical stratification occurs right at the ocean surface and is manifest in what is known as the microlayer. The microlayer has been variously described—from the top 50 micrometers (2×10^{-3} inches) to the top millimeter (4×10^{-2} inches). Functionally, it is the layer of water in direct contact with the atmosphere and therefore where airborne substances (including toxins) are dissolved and tend to collect in high concentrations— many times higher than in the water beneath. As these high concentrations suggest, this layer of water is cohesive and resists mixing with underlying waters.

Ecologically, the microlayer is home to a distinct flora and fauna, including plankton and bacteria along with eggs and larval stages of many species of invertebrates and fish. Fish fry, which hug the surface, also are often in direct contact with this layer. The full extent of the biological consequences of the high concentrations of toxins in the microlayer has yet to be documented, but widespread degradation has been hypothesized. Because the most sensitive life stages of many animals are spent here, it is likely that the impoverishment of the microlayer diminishes populations of adult animals living in deeper waters.

Zonation in the sea does not occur only with depth; there is also a pronounced latitudinal zonation, which appears to be related to major water masses. Currents form boundaries between water masses that differ considerably in their physical characteristics, and there is a predominant latitudinal pattern in the major ocean current systems of the world. The boundaries are neither impenetrable nor immovable, but distinct biological zones—with characteristic fauna and flora—are generally defined by the water masses. It may be related to major current patterns, which tend to be zonal with respect to latitude. Latitudinal zonation is also found on land, but it is more marked in marine species.[15]

Differences in Biochemistry of the Environments

Biochemistry is another reference point for comparisons between the diversity of life in the ocean and on land. Significant differences as well as similarities exist in the biosynthetic activities of marine and terrestrial plants and animals. The differences may reflect large taxonomic differences and/or may be related to the relative availability of certain elements in seawater as compared to terrestrial environments.[16]

Many marine plants and animals produce and release chemical substances that act as growth promoters or growth inhibitors or as toxins. These extracellular products are particularly common among the bacteria and algae. Toxins such as those produced by the red tide algae are well known because they contaminate seafood for human consumption. Perhaps even more interesting are the growth-promoting substances, which include chemical compounds required for the settlement and metamorphosis of many larvae. These substances are often produced by one species and trigger a response in another species, resulting in a mutualistic relationship between the two.

Abalone (a commercially harvested mollusk) requires a substance produced by a certain species of red alga, which is common on rocks in areas where abalone grow successfully. Without the red alga, abalone larvae would never settle and metamorphose to grow into adults. This mechanism guarantees that the abalone will grow only where there is a guaranteed food supply (the red alga). The alga species (which grows as a thin crust on the rocks) benefits, because taller algae, which would compete for space and light, are kept cleared away by the grazing abalone. In this way, the two species help maintain each other.[17] A similar relationship is known to exist between the sea hare and a red alga,[18] and between oysters and certain bacteria.[19] These chemical interdependencies between species may be quite common in the ocean.

Self-regulating chemical signals are also common among marine species. For example, mass spawning in beds of abalone is initiated by a hormone released into the water by some of the individuals. The hormone is the chemical signal, but hydrogen peroxide in the water will cause abalone to produce the hormone and therefore spawn. This fact is used to culture abalone, where hydrogen peroxide is added to culture water to stimulate the simultaneous release of eggs by the abalone. But imagine the chaos hydrogen peroxide pollution could cause in natural waters!

In fact, chemical interference may be a major threat to species that depend upon various biochemical stimuli in the water to control their reproduction, growth, and development. Anthropogenic pollutants could interfere with these mechanisms in such a major way as to threaten the existence of some species.

Gradients of Species Diversity

Several widespread patterns of diversity have been identified in the oceans. There are exceptions to all of them, and as more is learned they will undoubtedly be revised. One general pattern that has been both supported and disputed is that marine species (and terrestrial species) show a gradient in diversity with latitude—that is, species diversity increases with decreasing latitude. The diversity of habitats does increase along this latitudinal gradient (possibly because there is more area at low latitudes), so the total species diversity is likely higher for the tropics than for temperate latitudes. However, quantitive differences between tropical and temperate diversity on smaller scales seem to vary depending on the type of organism and type of ecosystem being compared.[20] It may also vary depending on which locations are chosen for the comparative studies and to what extent human pressures have already altered the faunal communities being studied. Several theories have arisen to explain natural differences in diversity between different ecosystems—theories based on factors that favor diversification.[21] Some of them would argue for greater diversity in the tropics while others simply offer possible explanations for differences in either direction.

- Tropical environments have been stable over longer periods of time, and therefore the evolution of species has simply gone on longer in these environments.

- Due to recent disturbances or newly created habitat, some ecosystems have not been open long enough for a full complement of species to be recruited from other areas.

- Certain ecosystems, such as coral reefs, are characterized by more physical complexity (called spatial heterogeneity), thus opening more niches for specialization by more species.

- Climatic stability generally results in higher diversity because species do not have to be as adaptable and tolerant of unexpected change; the tropics have less seasonal and long-term variability in climatic conditions, which has allowed the evolution of more specialized species.

- Even when temporal variations in climate do occur, if the patterns of change are predictable and cyclical, as in the case of seasonal cycles, species can evolve that adapt to or specialize on the temporal patterns.

- Increased competition maintains higher species diversity by preventing a single species from taking over the space, and if species in some ecosystems have finer or more overlapping habitat requirements, as they seem to, competition will be increased.

- Predation can maintain a higher species diversity by keeping down the populations of certain prey species that would otherwise exclude one or more weak competitor species.

- Continual removal of individual organisms by indiscriminate physical factors (e.g., storms, ice) can keep populations at reduced levels so that competition between species does not become exclusionary. Discriminate or highly lethal factors, however, tend to reduce species diversity (e.g., periodic anoxia).

- High productivity and primary production in marine environments may be associated with greater or lesser species diversity, apparently depending upon whether the productivity is sustained or episodic. The boom and bust cycles of primary production associated with fluctuations in nutrient and light conditions (as in highly productive ecosystems in high latitude environments and areas of periodic upwelling) seem to support a lower species diversity than does relatively stable primary production and sustained high productivity (as in coral reef systems).

The actual species diversity observed in an ecosystem will reflect the interaction of diversifying mechanisms and other factors that degrade ecosystems and reduce diversity, such as chemical pollution; thus latitudinal gradients may or may not be confirmed.

Another pattern of importance is that species diversity tends to exhibit a positive gradient moving off shore. Among planktonic species, offshore waters have a much higher diversity than inshore areas. Similarly, diversity of benthic species increases with depth: from the more productive shelf regions out to the low-productivity bathyl regions and at least to upper abyssal depths. For example, off the Atlantic coast of North America, species diversity increases with depth to a maximum at around 1,800 meters, or possibly at 3,500 meters.[22] The same peak of diversity may not be found in the Pacific Ocean, where deep-sea species diversity at 5,000 to 6,000 meters is apparently as high as at shallower depths.

Another interesting geographical pattern is found in species diversity of coral reefs. There is a longitudinal diversity gradient decreasing from west to east in both oceans, and diversity tends to be greater north of the equator. Also, Pacific coral reefs are on the whole more diverse than Atlantic reefs. This is consistent with the tendency for many types of plants and animals to be represented by a higher species diversity overall in the Pacific Ocean than in the Atlantic.

Eastern Pacific reefs tend to be less diverse than western Atlantic reefs. Although mollusks exhibit a west to east decrease in species diversity across both oceans, coral species diversity in the western Atlantic is greater than that in the eastern Pacific. This may have to do with a greater sensitivity by corals to negative effects of the eastern Pacific upwelling areas.

Biological Diversity in Different Types of Marine Ecosystems

When comparing factors controlling the biological diversity of the different types of ecosystems, it is important to distinguish between biotic and physical control mechanisms in an ecosystem, and marine science lags behind terrestrial science.[23] An interesting relationship between the physical environment and benthic species diversity has been noted: where the physical environment is less limiting (i.e., less severe), there tends to be higher species diversity controlled primarily by biological interactions.[24] Conversely, where the physical environment is severe, it controls the biological community, which generally has a lower species diversity. This relationship may help explain some of the natural gradients of species that were described previously.

Marine biological communities may be divided into two broad categories: pelagic and benthic. Pelagic communities live in the water column and have little or no association with the bottom, while benthic communities live within, upon, or associated with the bottom.

The organisms living in pelagic communities may be drifters (plankton) or swimmers (nekton). The plankton includes larvae of benthic species, so a pelagic species in one ecosystem may be a benthic species in another. The great abundance of primitive organisms such as bacteria (bacterioplankton) and viruses (femtoplankton) has only recently been appreciated,[25] and their ecology and genetics are poorly understood. It has been suggested that the viruses are important in the exchange of genetic information among the phytoplankton. Because so little is known, we cannot adequately discuss these primitive organisms, let alone measure their contribution to the genetic, species, or functional diversity of the marine environment. It is important to note, however, that micro-organisms are significant players on the biological diversity stage in the sea.

Pelagic communities include the loosely organized assemblages within coastal boundary current systems, which are structured by unpredictable advective and upwelling processes, and the well-organized, self-regulated assemblages in vast cyclonic (counterclockwise) or anticyclonic (clockwise) gyres. The latter communities tend to be predictable, and they characterize dominant circulation patterns in the world oceans.

Pelagic systems are thought to be controlled primarily by physical factors, including temperature, nutrients, light in the surface waters, and disturbances in the water structure. The latter occurs when winds and other atmospheric conditions drive changes in the circulation patterns of ocean waters. As a result, there are vertical changes in the temperature and nutrient distribution, which in turn affect the vertical distribution of species. There is no clear evidence of biological factors controlling species diversity in these ecosystems, but species interactions have not been well-studied.[26]

The greatest known diversity of marine species exists in benthic communities, especially in coral reefs and on the deep-ocean floor. The benthic environment includes the intertidal shore, the shallow subtidal or continental shelf, the continental slope, the deep abyssal plains, and isolated ecosystems such as coral reefs, seamounts, and deep-sea trenches. The substrate may vary considerably, with distinct differences between hard-bottom and soft-bottom communities. Type of bottom has a pronounced effect on the nature of the community that lives there. Beyond that single physical factor, species diversity is maintained by biological mechanisms—competition, predation, larval recruitment, and biological structuring of the substrate—and/or physical mechanisms, such as nutrients, light, waves, and currents.

Benthic habitats are often classified as hard bottom and soft bottom. According to one theory, processes that tend to increase diversity in hard-bottom communities are most often controlled by physical or biological disturbances, which moderate competition between species to the point of preventing exclusion of the less successful competitors.[27] For example, if repeated disturbance (e.g., storms or predation) causes continued or intermittent removal of part of the population of a species that would otherwise take over because of its highly successful competitive strategies, there is space for other less successful competitors to survive. On the other hand, increased diversity in soft-bottom communities is usually the result of biological activity, which adds heterogeneity to the environment, thus creating new niches for new species.

Biomes and Biogeographical Provinces

There have been few attempts to catalog the various biotic systems of the world's oceans. In fact, the term *biome* is remarkably absent from marine ecological literature, despite the fact that it is a basic unit used by terrestrial ecologists. A biome is a characteristic type of community of plants and animals in a characteristic type of natural environment. Several regions worldwide may belong to the same biome. On land, these distinctive communities have evolved mainly in response to dramatic variations in climate and topography. (The tropical rain forest is one of about a dozen major terrestrial biomes.) Several coastal and ocean biomes can be identified, and they are discussed fully in Chapters 3 and 4.

Biomes are ecologically defined, but if a geographical component is added, a biogeographical classification system results. Various systems of realms and regions (defined by physical environments and geographic boundaries) and provinces (defined by typical assemblages of biota) have been proposed for the marine environment.[28] Unlike biomes, these systems do not include the third dimension—the depth of the ocean—and therefore much of the ocean biota is left out.

Chapter Three ▄▄▄▄▄▄▄

Marine Ecosystems: Coastal

FROM A PURELY ecological perspective, marine ecosystems most naturally fall into the categories of benthic and pelagic. However, for the policy maker and the environmental manager, it is more natural to divide the ocean into the waters that are within governmental jurisdictions and are interconnected with the land and its human societies and the waters that are remote and are part of the global commons under no particular jurisdiction but subject to international law. Therefore, it is more appropriate for the purposes of this book to consider the coastal and the oceanic realms separately. This chapter and the next identify the major biomes (ecosystem types) in each realm and characteristics of the biological diversity and inherent processes maintaining diversity in each.

The coastal zone is characterized by interconnections among neighboring ecosystems, and it is directly influenced by the land and land-based human activities. The coastal zone includes the land-sea-air interface zone around continents and islands and is defined as extending from the inland limit of tidal or sea-spray influence to the outer extent of the continental shelf. Within these

41

limits lies a multitude of ecosystems, which, because of the nature of water, interact extensively with each other and with neighboring terrestrial ecosystems. It may also be argued that the inland boundary of the coastal zone extends up the freshwater streams that drain into marine ecosystems, because the impact of these waters and of their loads of sediment and dissolved materials is so great that they too are part of the interconnected coastal ecosystems.

Impacts on one coastal ecosystem may spill over into another, following water circulation patterns or traveling through marine food chains. The exchange is not only chemical and physical, for organisms can also migrate between ecosystems. Biological dispersal is a critical aspect of marine ecology, and the fluid environment often allows for broad distributions of species. It is not unusual for a marine species to spend one life stage in one ecosystem and the next life stage in another. There are also animals, such as the numerous species of seabirds, that use both terrestrial and marine environments.

Coastal marine biomes include a wide variety of discrete but interacting ecosystems ranging from estuaries and inland seas to the depths of the outer continental shelf. The boundaries of these ecosystems may include hard physical barriers such as basin or coastline topography and soft structural barriers such as water currents or vertical stratification. Often an additional type of boundary is introduced by the humans who use and regulate the ecosystems—the political boundary. This includes national borders, the limits of territorial seas, and the exclusive economic zones (EEZs). These borders are convenient and often essential for managers, but it should be kept in mind that circulating waters do not honor political boundaries, and therefore marine ecosystems cannot be effectively managed by fragmentary regulations. Integrated, whole-system approaches are required, not only for single ecosystems but often for the entire coastal zone.

Throughout this discussion of diversity in the different marine biomes, it is important to keep in mind that a higher species diversity does not impart a greater value to one biome over another. Characteristic species diversity does not necessarily reflect the complexity or value of biological functions in the ecosystem. What is important is that an ecosystem maintain a characteristic species diversity that enables the maintenance of important ecological functions.

Coastal Benthic

Coastal benthic ecosystems are found wherever land meets sea, and they are the most vulnerable to the negative effects of economic development and burgeoning human populations. It may well be that coastal benthic ecosystems are as much at risk as the tropical rain forests of the world. Certainly, many scientists and environmentalists recently have claimed this to be true for coral reefs.

Rocky Intertidal and Subtidal Shores

Rocky intertidal ecosystems are moderately diverse systems, with apparently higher diversity of animal species in tropical regions than in temperate, but higher algal diversity in temperate areas. There is usually a clear pattern of zonation in species distribution with respect to tidal height. More species are present in lower zones, where the physical environment is less stressful, than near the high-tide level, where periods of exposure are greatest. Seaweeds flourish in the rocky intertidal and the shallow subtidal regions just below low tide. This is the biome in which seaweed diversity is greatest, because the requirements for light and a firm substrate for attachment are both met.

The species composition of intertidal zones is controlled by several factors: (a) physical factors such as desiccation, high light intensity, and large temperature fluctuations during exposed periods; (b) physical factors affecting food supply; (c) success of recolonization; (d) grazing and predation; and (e) competition. The relative importance of these factors varies from one location to another.[1]

Nevertheless, several researchers have sought to identify dominant factors controlling species diversity. One theory proposes that diversity is controlled by the presence of predators and the efficiency with which they prevent monopolization of space by competitively dominant species.[2] Comparisons of subtropical and temperate communities reveal a higher ratio of top-level predators to total species in the subtropical community, which also has greater total species richness.

Several researchers have suggested that predation is usually the dominant factor controlling species diversity. In the presence of predation, competing species can coexist even when the competition is unbalanced. Predation on the superior species reduces the competition and allows the inferior species to survive. For example, when predators (starfish) were removed from a temperate rocky intertidal area, diversity declined as half the prey species disappeared. The space was taken over by the competitively superior species (mussels). Furthermore, once the mussels grew to a big enough size, reintroduction of the predators had no effect because they could not eat the large mussels.[3]

This study demonstrates how a disturbance causing the loss of one element in a community can have a cascading effect, reducing species diversity even more. Furthermore, elapsed time may be a critical factor in determining whether a disrupted community can recover its characteristic diversity once the initial disturbance is eliminated. This has major implications for the regulation of human activities in coastal areas and the potential effectiveness of recovery programs.

While the predation theory seems to hold well for diversity of species near the bottom of the food chain (these species are competing primarily for space), some ecologists argue that competition has an increasingly dominant role in maintaining species diversity at higher trophic levels (these species compete mainly for food). The practical effect is the same, however: removing one species may ultimately cause several others to disappear.

Two additional factors contributing to species diversity in rocky intertidal systems are physical disturbances and recruitment of larvae.[4] Benthic animals produce planktonic larvae, which drift on currents, often far from the site of their release. To maintain a position of significance in a community, the species has to be replenished by settlement of larvae. Populations therefore can fluctuate considerably in response to success or failure of larval recruitment. Thus, environmental conditions affecting the larvae also have an effect on the benthic community, with an ultimate effect on the species diversity. Algal replenishment occurs locally or from distant populations by spores or drifting pieces.

Rocky intertidal animal communities generally follow the gradient of species diversity increasing with decreasing latitude. It is commonly assumed that tropical species are more specialized, which means that each has a smaller niche (role in the community) and so more species can share the ecosystem. A study of marine snails in temperate and tropical intertidal areas suggests, however, that tropical species are not necessarily more specialized,[5] and that some other explanation must be sought for the greater species diversity at low latitudes. Equal numbers of habitat generalist species were found at both latitudes, but there were many tropical niches (types of jobs, so to speak) with no temperate counterparts. Thus, the reason for increased species diversity in the tropics, it was proposed, is the greater diversity of niches that can be occupied.

A special situation found in some rocky intertidal areas is heavy surf action, which causes the organisms to be battered constantly by waves. Many species cannot hold on under these extreme physical conditions, but those that can are unusually productive, composing possibly one of the most productive ecosystems known. Wave energy indirectly enhances productivity on intertidal wave-beaten shores by washing nutrients and suspended food to sessile plants and animals and by keeping seaweeds wet during exposed periods so that rates of photosynthesis are maximized. Waves also wash unattached grazers and predators away or make it difficult for them to feed. Although diversity is limited in this highly disturbed area, certain species that could not compete well elsewhere thrive here, thereby increasing diversity in the intertidal zone as a whole.

Rocky intertidal areas are subject to overharvesting, pollution due to runoff and waste outfall from human development, and habitat alteration. The threat will continue to grow with the burgeoning human population and its tendency to concentrate near the shore. The tides and waves bring in protective water laden with food but often also laced with harmful heavy metals, petroleum products, or toxic organics. Some animals and plants are lost: some slip off the rocks, some become diseased or deformed or die, some cannot reproduce successfully. Larvae may not find the proper environment or the requisite chemical signal to make them settle, so populations are not replenished. Eventually, some species disappear and others lose genetic variation within their populations. The characteristic biological diversity is diminished; the system is weakened and becomes more vulnerable to further stresses.

Dramatic climatic change, such as that predicted to occur as a result of air pollution and the greenhouse effect, may cause major shifts or significant losses in species diversity. As discussed at a 1988 symposium on climate change:

> In the oceans, the survival of a species often depends on the dispersal of larvae to new places. Many invertebrates produce mobile larvae, which settle and continue their development only where the conditions are right. Changing climate and [the resulting] changing patterns of currents in the oceans could condemn many to stay in their larval form, never finding a place to settle and eventually dying without producing offspring.[6]

Sandy Shores and Mud Flats

Both sandy shores and mud flats are low-diversity systems that include specialized species not typically found in rocky intertidal areas. Both environments are characterized by lack of attachment sites for animals and by limited food availability.

The sandy intertidal environment is physically unstable because of shifting sands, and therefore few species successfully establish themselves there. In surf zones, the fauna are limited to species such as the razor clam, which can escape low tide and shifting sands by quickly burrowing, or the pismo clam, which has a very thick and smooth shell to withstand the battering of the surf and sand. Food species in this environment are also few and highly adapted—primarily, a few species of diatoms, which live only in the surf along sandy beaches. They are highly productive and produce a reliable food source for razor clams in intertidal sands. As the tide recedes, the floating masses of diatoms are left on the sand surface and then are resuspended on the incoming tide.[7]

Mud flats do not experience the physical disturbances characteristic of the surf zones, but respiration is a problem in the oxygen-poor muds. Animal species are dominated by worms, clams, and snails. The fiddler crab is a species specially adapted to this environment. Large concentrations of them build airtight burrows in the mud, where they spend high tide. The burrow has a "plug" in the door and during low tide the crabs "open the door" and come out onto the mud flat to feed on organic material left by the tide.

These ecosystems are threatened by the same environmental hazards as are other intertidal areas. The species, although few, tend to be tolerant of sharp changes in the environment because that is what they deal with on a daily basis. It depends on the nature of the changes, however; low oxygen levels, for instance, are not problematic to this community, while toxins may be.

Estuaries and Wetlands

Estuaries and associated wetlands lie at the fringes of the marine environment—inlets, embayments, river mouths, lagoons, and small enclosed seas—where influences of inflowing river water as well as ocean tides are at play. These ecosystems are individually characterized by a relatively low species diversity, but they are functionally complex. Collectively, estuaries around the world support an important diversity of plant and animal species, some of which include early life stages of migratory species. In fact, much of the biological production of an estuary or wetland is carried out of that system by tides or migrating animals.

Estuaries and wetlands have been extensively studied because of their high productivity, their importance to commercial and sport fisheries (and "game" birds), and their use and abuse by coastal developers. The species diversity of most of these systems is well known because it is relatively easy to study. Most ecological studies have focused on the productivity and the variety of biological processes occurring in these ecosystems. There has also been some discussion about the processes that control species diversity.

As is generally the case with low diversity ecosystems, the physical environment dominates the biology. It has been suggested that the low species diversity characteristic of estuaries can be attributed to the relative inconsistency and unpredictability of the physical environment.[8] There are generally large gradients of salinity, which may fluctuate with the tides, and at middle to high latitudes there are large seasonal variations in temperature and light. In addition, there may be periods of anoxia as well as fluxes of suspended sediments. The fluctuating environmental conditions lead to physiological stress, so the species that do well in these systems must be highly adaptable to varying physical conditions. It is, however, important to emphasize that the low diversity does not imply lack of stability. On the contrary, estuarine systems are considered highly resilient and resistant and therefore stable.

Another factor hypothesized to explain the low species diversity in estuaries is their relatively short geological life span. Due to waxing and waning sea levels and ice ages and the drift of continents, estuaries tend to be quite young geologically, so that they have not had enough time to establish more complex communities.[9]

A consistent gradient of species diversity is associated with the salinity gradient in estuaries. Diversity declines progressively from the coastal sea into less saline (brackish) waters. Most of the decline is due to the disappearance of whole families of species. Once fresh water is reached, the diversity increases again. A few estuaries have salinities higher than typical ocean waters, and these also tend to support lower species diversity.

Salt marshes, mangroves, and eelgrass beds may be viewed as special kinds of estuaries, each dominated by seed-producing, vascular plants specially adapted to living partially or entirely submerged in shallow saline waters. They are often located along the edges of large estuaries such as bays and inland seas. The dominant plants are more highly evolved than algae and come from plants more commonly terrestrial. In addition to the one or few species of dominant vascular plants, there may be many species of attached or drifting algae, which contribute significantly to the species diversity. There may be hundreds of species of algae associated with salt marshes, but they do not support a diverse consumer fauna.[10]

Among the most important characteristics of the various types of estuarine biomes is their role as nursery grounds for many species of coastal and oceanic fish and shellfish. Consequently, environmental changes that affect larvae and juveniles here can greatly impact the populations and distribution of species found as adults in other biomes.

As a result of their shallow waters and restricted circulation and the coastal developments often associated with them, these ecosystems are highly susceptible to environmental degradation from human activities. Habitat destruction from channelization, drainage, filling, and water diversion is not uncommon. In addition, eutrophication and toxic pollution have become extreme in many cases, and turbidity is increased with the influx of suspended sediments from erosion and construction along inflowing rivers and on the estuaries themselves. These disturbances generally limit species diversity and reduce ecosystem stability. Many salt marshes, mangroves, and eelgrass beds have permanently disappeared because of human activities, and countless estuaries are impoverished or dying.

One of the important issues concerning estuaries and wetlands is that of mitigation. Can the impact of human activities on these sensitive ecosystems be reduced by efforts to restore them to their characteristic diversity of species and functions? Once damage has been done, simply stopping the offending activities may not be enough to bring the system back to its original state.

Another impact on biological diversity that should be mentioned in this discussion of ecosystems at the transition between land and sea is the potential effect of marine pollution and climatic change on animals that feed but do not live in marine waters. Many species of shore birds fall into this category, as do mammals such as river otters, mink, and raccoons. Perhaps most at risk are the coastal migratory birds, which travel long distances and depend upon ample food supplies along the way. For example:

> Shorebirds such as sanderling and plovers spend the winter in South America and travel north to breed in the Arctic in summer. They must arrive in the north and hatch their young just as the arctic insects undergo their summer population explosion, providing the young with ample food. The adults must also feed on the journey, however, and every year they stop in Delaware Bay to eat the eggs of horseshoe crabs. The crabs arrive and lay their eggs at the same time each year. If these events get out of phase, the effects on the migrants would be catastrophic.[11]

This type of finely tuned synchrony can easily run amok if rapid environmental changes cause dramatic deviations in the food supply.

Although not as tied to a precise schedule, birds that nest along

.he shore, particularly colonial birds, can be dramatically affected by changes in the food supply. Changes in species diversity of shore birds can be a good indicator of biological changes (e.g., changes in biological diversity and productivity) occurring in the coastal marine ecosystem.

Coral Reefs

Coral reefs have the greatest species diversity of any marine biome described to date (although it may turn out that the deep-sea floor has a greater diversity). Many people are more familiar with the diversity of these systems because of the relative accessibility and inherent beauty of the biological communities—spectacular photographs have been taken of the colorful species in these clear waters. The densely packed variety of colors and forms of life in these ecosystems are cause for awe and admiration to all who view them. The diversity of animal species is similar to that in tropical rain forests. The plant species, however, are not as diverse. Also similar to tropical rain forests and unusual for marine environments is the characteristic physical structure mentioned in Chapter 2.

John Wells wrote in 1957, "Reefs are scattered over an area of 190,000,000 square kilometers (68,000,000 square miles) wherever a suitable substratum lies within the lighted waters of the tropics beyond the influence of continental sediments, and away from the cool upwellings of the sea in the eastern parts of the ocean basins."[12] Light is required because the corals harbor unicellular algae that live symbiotically within the living coral. The algae photosynthesize, producing organic material that is passed into the coral tissue, providing nutrition. Thus, the animal does not need to capture food from the water. Consequently, although the coral is an animal, its growth is as dependent upon light as a plant's.

The calcium carbonate structure of reefs is composed primarily of the skeletons of coral; living corals exist only at the outermost surface of the structure. Calcareous algae also help in producing coral reefs—sometimes they cement together the corals, sometimes they form their own calcium carbonate masses over the top of corals. Different species of coral produce characteristically shaped calcareous "homes"—thus, the staghorn coral, the brain coral, and so on. The base, or dead part, of a reef is a composite of skeletons of the multitude of coral and calcareous alga species that have been associated with that reef over time.

The general setting—the structural complexity, the low level of environmental fluctuations, the clear water, the age—is conducive to the development of a complex community with a high species richness. It is widely accepted that the species assemblages are highly organized and have coevolved, and that it is competition that has given rise to the large numbers of specialized species. The natural diversity of any particular coral reef is determined by intermittent natural disturbances and/or chance colonization as well as by its geographical location. As previously mentioned, there is an interesting gradient in species diversity, increasing from west to east in both the Atlantic and Pacific oceans.

Coral reefs represent an unusual type of benthic environment, one with a great deal of structural variability because of the calcium carbonate structure of the coral itself. One theory suggests that diversity in coral reefs, as in tropical rain forests, is maintained at a high level by natural disturbances rather than by environmental constancy.[13] Examples of such disturbances on reefs are storms, freshwater floods, sediments, invading herds of predators, and perhaps gradual climatic change. This theory depends upon unbalanced competition: one of the competitors is always more successful and would eventually exclude the other were it not for the external disturbances that cause greater mortality in the more successful of the pair. In the absence of disturbance, diversity would be maintained at a lower level. Keystone predators are important disturbers in reef communities.

Others have argued that it is the interaction of disturbance and growth rates that determines diversity.[14] For instance, there is an increase in diversity of coral species with depth on a coral reef (down to a critical depth at which diversity is severely limited by too little light). It has been suggested that very rapid growth rates in higher light intensity outstrip the effects of disturbance, resulting in competitive exclusion of the weaker species; factors of disturbance at greater depths dominate. This is a possible explanation for the increase in species diversity with depth.

It has been noted that, on coral reefs, fish species assemblages are not stable—if disturbed, fish species do not necessarily recolonize in the same assemblage. This has led some to conclude that chance colonization may often be the best explanation for the diversity found on a specific reef.[15] Others, however, have associated diversity with water temperature, noting that the highest diversity is associated with the warmest surface water temperatures.[16] They claim that evolution has occurred more rapidly in warmer waters. Yet another theory associates nutrients with diversity: the greatest diversity is found in nutrient-poor waters.[17]

Keystone predators may play a significant role on coral reefs. A prime example is the crown of thorns starfish on the Great Barrier Reef and other Pacific reefs. Fluctuations in the population of that efficient predator can cause dramatic differences in species diversity of reefs. During heavy infestations, large portions of reefs are destroyed, but they seem to be able to recover after the plague is gone—depending on how extensive the damage is. There has been some speculation as to whether pollution promotes population explosions of the crown of thorns, but the majority of experts seem to believe that there is no connection.

On Atlantic coral reefs, the long-spined sea urchin *Diadema* is a prominent keystone predator. A massive die-off of the population in the Caribbean in recent years had little effect on the rest of the community, but it was finally concluded that so many *Diadema* were still left that they could maintain the species' critical reef role. This demonstrates how complicated the community structure is and how difficult it is to identify the critical components to use as indicators of change in community diversity and/or stability.

It should be mentioned that coral reefs are not the only biologically formed reefs in the ocean. Certain species of tube worms (worms that secrete and live in shell-like tubes) also form contiguous calcified structures when their tubes are cemented together, and calcareous algae can build reef structures. These become the physical framework for diverse communities similar to the communities on coral reefs. Studies of these ecosystems in temperate and tropical waters indicate that the tropical environment is not necessarily more stable than the temperate, and intermittent disturbances best explain the greater diversity found on tropical polychaete (worm) reefs.

Referring to the near-crisis proportions of biological degradation among the world's tropical marine ecosystems, especially coral reefs, John Ogden has pointed out that "the preservation of biological diversity is not primarily a scientific issue but rather a social one. Declines in species-rich marine environments are the result of exploitation of resources, in most cases by indigenous human society with immediate economic or food needs and no alternative."[18]

Direct human threats to coral reef diversity have been and continue to be in the form of overharvesting and destructive fishing. Because of these activities alone, the coral reefs of the world are in a critical state. There are also increasing threats from nutrient enrichment and siltation from agricultural runoff, urban development, mining activities, and deforestation of rain forests. Corals and their symbionts require clear waters for optimal growth. Additional nutrients from runoff promote the growth of planktonic algae, which in turn shade the corals and reduce the photosynthesis of their symbionts. Increased sediment loads washed from land in association with mining and deforestation also shade the corals, and, even more important, they tend to settle on the corals and smother them. Chronic oil pollution can significantly lower species diversity in areas where there are refineries, drilling, and tanker traffic.

As if that isn't enough, the effects of global warming may devastate many reefs. Corals in many tropical waters today are living very near the maximum of their narrow temperature tolerance range, so a small increase in temperature from global warming may be lethal to coral reefs regionally if not globally. Trouble has already been noted. For several years, widespread coral bleaching—a stress phenomenon in which corals expel their algae during periods of physiological duress—has occurred during the warmest season in reefs worldwide. Secondarily, sea level rise, if rapid enough, could outpace coral growth leaving some reefs too deep to survive.

Subtidal Continental Shelf

Within the coastal zone, but deeper than the coastal fringe ecosystems, lies a great expanse of continental shelf that supports a benthic fauna with a low to moderate species diversity. This environment generally falls within the depth range of about 20 meters to about one hundred meters at the edge of the continental slope, which drops precipitously to the deep abyssal plains of the central oceans. Generally speaking, this is a soft-bottom environment with insufficient light to support much photosynthesis. Consequently, with the exception of isolated giant kelp forests, which reach upward to well-lighted waters, the benthic communities of the continental shelf are dominated by animals.

The diversity of benthic habitats is greatest in the tropics. Individual types of communities or types of fauna (e.g., sediment infauna or sediment surface fauna), however, may be more or less diverse than their temperate counterparts.[19] Though usually these shelf communities are soft-bottom communities, here and there rock outcroppings break through, providing habitat for those species preferring hard bottoms. Sometimes these hard-bottom areas support an unusual and diverse community. A good example is Cordell Bank, one of the U.S. National Marine Sanctuaries.

Vertical currents known as upwellings can create low oxygen conditions in sediments underlying these areas, with low diversity benthic communities dominated by only a few species. The upwelling waters are rich in nutrients, support a high productivity near the surface, and are characterized by high oxygen demand. Because the periods of highest production of microalgae are followed by oxygen depletion when the unconsumed algae die and decay, the sediments become anoxic, resulting in low animal populations and low species diversity and extensive microbial mats.

The continental shelf off the northeast coast of the United States is one of the most intensely studied regions in the North Atlantic, and may be used as grounds for comparison with the deep-sea floor of the North Atlantic. About 500–600 species have been identified in benthic communities, and this probably includes much of what is there. The species fall into 13 phyla, but only 4 of these dominate the faunal composition. Density and biomass (total weight) of animals in these productive waters is much greater than in the deep sea. The distribution of animals is well correlated with the different types of sediments.[20]

It is not difficult to imagine that human activities may affect the benthic ecosystems of the continental shelf. Perhaps one of the major threats is in the form of contaminated sediments. Industrial and urban outfall into rivers, estuaries, and open coastal waters and barge-dumping of dredge spoils, sewage sludge, toxic industrial wastes, and radionuclides contaminate both water and sediments. Large quantities of dredged sediments from contaminated harbors, and drilling muds from offshore oil and gas exploration, are dumped in coastal waters at selected sites. All this toxic material bound in sediments is a direct threat to the health and survival of benthic species, some of which burrow and feed in the sediments. All are at least in contact with them.

The typical pattern of community response to toxic sediments is a loss of diversity, with the survival and proliferation of a few tolerant or opportunistic species, while less tolerant species become rare or extinct. For example, diversity gradients in benthic communities in Norwegian fjords have been attributed to the distribution of pollution.[21]

One specialized subtidal community with relatively high diversity is the giant kelp bed. As mentioned, it is uncharacteristic in that it is dominated by plants. The kelp are so long (up to 100 meters) that they can float their photosynthetic leaflike fronds near the surface while anchoring themselves on the bottom. This physical structure adds a greater diversity of niches and therefore supports a diverse community. As with coral communities, damage to the living structure—in this case, the kelp—will alter the whole community. They can be reduced significantly by overgrazing, so a balance between kelp, grazers, and predators on the grazers is important in maintaining species diversity in this ecosystem. The extent and distribution of offshore kelp beds have been significantly altered by changes in the populations of keystone predators.

Coastal Pelagic

Coastal shelf waters, with maximum depths on the outer shelf of about 1,000 meters, exhibit less stable circulation patterns than deep-ocean waters and, consequently, the environment fluctuates more. Coastal shelf waters are characterized by strong current regimes and zones of periodic upwelling, where bottom waters rich in nutrients for plankton growth move to the surface to replenish waters carried offshore by wind-driven currents. These rich areas have a high productivity but a relatively low species diversity, consistent with the gradient theories discussed previously.

Such areas are often the sites of major fisheries, since schools of fish congregate in these waters to make use of the plentiful food supply. At least 80 percent of commercially important marine species live within 200 miles of coasts. Fluctuations in currents and upwellings cause significant variations in productivity of the system and, therefore, in the yield to commercial fisheries.

The coastal pelagic food web includes phytoplankton, zooplankton, larvae, fish (representing several trophic levels), marine mammals, seabirds, humans, and, of course, bacteria. The energy base of these highly productive ecosystems is provided by phytoplankton in the photic zone, which respond to seasonal and interannual oscillations in nutrient supply. When nutrient-rich waters rise to the surface, phytoplankton production surges—particularly when the nutrient enrichment is timed with favorable light conditions. There is some loss of this primary production to the decomposition cycle, when algal growth in excess of grazing dies off and sinks. However, zooplankton populations respond to the increased algal production with their own population surges. These increases in productivity are passed along to the fish food web, yielding large populations of a relatively low number of species. Schooling fish are favored by this type of nutritional environment, the schools following patches of upwelling and high production. The bottom waters of upwelling areas, however, are often depauperate, both in total populations and in number of species. The high oxygen demand of the productive waters above may deplete the oxygen in bottom waters, creating an unfavorable environment for most species.

The pelagic ecosystems in deep ocean or shelf waters are particularly interesting in their structure because there are no impenetrable physical barriers defining them. The barriers to dispersal of plankton species in these ecosystems are subtle and not entirely understood. Some of the more obvious barriers include strong and persistent hydrographic features such as fronts (currents), rings and gyres (cyclonic or anticyclonic movement of water masses), pycnoclines (sharp vertical thermal and salinity gradients), and vertical pressure gradients.[22]

In addition to these subtle physical barriers, there are biological limitations to dispersal, such as limits on the distance an individual may travel and still be integrated into a new population or inhibitions to interbreeding among different populations of the same species. These are difficult to determine, but they do exist and must be taken into account when considering species and genetic diversity. Genetic analyses of zooplankton and phytoplankton populations evidence spatial and temporal isolation of genetic groupings.

Zooplankton, for example, are distributed in distinct communities that can be defined geographically and seasonally in waters overlying the continental shelf. For example, off the northeast United States, coherent ecosystems are defined by characteristic species assemblages of zooplankton in the Gulf of Maine, on Georges Bank, and in the Mid-Atlantic Bight.[23] Little variation has occurred over several decades, suggesting definition of separate ecosystems with differing physical properties of the water masses.

Estimates of predation on fish in one rich fishing ground, Georges Bank off the coast of New England and Nova Scotia, indicate that 60–90 percent of the fish production is consumed by other fish, while 8–14 percent is consumed by humans and only slightly less than that by marine mammals.[24] Species diversity— and, especially, relative abundance of species—in these ecosystems can vary considerably in response to interannual variations in climate, currents, upwelling patterns, fishing pressure, and predation dynamics.

The regulation of fisheries to prevent overfishing presents a considerable challenge because of all the variables involved. Recently, fisheries managers have recognized the advisability of managing large marine ecosystems instead of individual fisheries species. The theoretical goal is to establish fishing regulations (perhaps with annual variations) to attain the maximum sustainable yield of target fisheries species. In order to do this, it is essential first to know what populations of the different species are sustainable, and then what amount of fishing will allow those populations to be maintained. Because many species besides the target species are harvested during fishing operations, and because fishing can have indirect as well as direct effects on populations by affecting competition and predation dynamics, it is now clear that the whole ecosystem must be considered in any fisheries management plan.

Georges Bank provides an example of the importance of interactions between species—in this case, the commercially important herring and mackerel, and the sand eel, which is not fished. Herring prey on sand eels, sand eels prey on herring larvae, and mackerel prey on larvae of both the others. Heavy fishing pressure on the herring and mackerel allowed the sand eels to flourish and, in fact, attain population dominance over the previously dominant herring. The predation by herring and mackerel was reduced, and the sand eel compounded the imbalance by preying more heavily on the herring larvae. Only by significantly reducing the fishing pressure on the two commercial species can the populations of the three species be brought back into balance.

Coastal Basins

Enclosed or semiclosed seas are coastal basins that fall somewhere between huge estuaries and small oceans. Examples of such seas include the Gulf of Mexico, the Mediterranean Sea, the Baltic Sea, the Bay of Bengal, the South China Sea, and the Sea of Japan. They usually support major fisheries and are bounded by several nations, so they are best managed by regional international agreements. These seas usually include several of the biome types discussed in this chapter—e.g., wetlands, estuaries, continental shelf, coastal pelagic, and deep-sea floor. Environmental damage from pollution is intensified due to the restricted circulation and exchange with outside waters. The severely degraded state of the Mediterranean ecosystems is a good example of this.

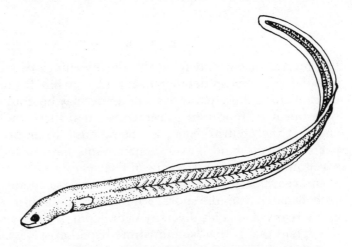

Chapter Four

Marine Ecosystems: Oceanic

T H E open-ocean biomes are defined vertically by depth and horizontally by major permanent currents. The currents form boundaries between water masses that have distinct physical, chemical, and biological properties. In addition to this horizontal distribution of discrete water masses, there is a vertical layering that coincides with sharp gradients in physical properties of the water column—temperature, salinity, light, oxygen, and minerals. The resulting parcels of water—which are wrapped around the sides by currents and bounded on the tops and bottoms by more or less stable boundaries—may be thought of as separate ecosystems, each with characteristic biological communities. However, both the currents and the vertical gradients represent "soft" boundaries that allow migration of organisms, detritus, and dissolved materials between ecosystems. The uppermost boundary is the air-sea interface, and the deepest boundary is the sea floor, which is an ecosystem unto itself—one that supports a fascinating diversity of life.

The open ocean is also subject to influences from land—through the air and water currents that move from the coastal region into the central ocean and from the downwellings of surface waters into the great ocean depths. It may take a while for material to be carried along these pathways, and there may be dilution along the way, but it is important to remember that these routes are open and that the central ocean is not immune to anthropogenic impacts from the land. Toxic contaminants have in fact been measured in the tissue of deep-ocean fish species.[1]

As described in Chapter 2, species diversity tends to increase with distance from the shore. For benthic communities, this means an increase of species diversity with depth. A theory that attempts to explain this is the "stability-time hypothesis,"[2] which proposes that diversity is highly correlated with the physical stability and length of past history (geologic time) of an environment. Deep-sea climate does not fluctuate, while the waters of the continental margins are subject to seasonal and unpredictable changes. The shallower waters have also been affected by long-term changes in global climate and sea level and even by the longer-term movement of continents. Many coastal ecosystems are formed, destroyed, or re-formed during such events.

Deep-sea environments that do not have the prerequisite large-scale stability over time and space have a greatly reduced species diversity. For example, deep-sea trenches are prone to frequent large slides down their steep walls; in areas of upwelling bottom waters are constantly or intermittently low in oxygen; and areas of strong bottom currents have bottom sediments that are continually stirred up and moved along current paths.

It is somewhat more difficult to assess the diversity gradient of the pelagic community with distance from shore, because the vertical stratification into layers of distinct communities may not be comparable. Nevertheless, near-surface communities, at least, seem to follow the rule that diversity increases with distance from shore.

Deep-Sea Benthic

Perhaps the most interesting story is in the deep-sea benthic community, which lives a thousand to several thousand meters beneath ocean waters. The first scientists to study the ocean floor envisioned a hostile environment almost devoid of life. They reasoned that the high pressure, cold temperature, and low food supply would allow very few species to survive. Early attempts to sample the deep-sea fauna seemed to confirm these ideas, but, unbeknownst to the researchers, most of the specimens collected from the sea floor were lost on the long trip up to the surface.

Improved sampling techniques in the late sixties allowed scientists for the first time to recognize the high species diversity of the animals living in deep-sea sediments in the North Atlantic. Although they found total biomass to be low, both the number of species and the evenness—because of the homogeneous environment—were high. The fact that many of the more abundant species were broadly distributed indicated that barriers between deep-sea communities are less distinct than in shallow water.[3] The high species diversity is associated with low productivity.

The deep-ocean environment has been stable over long periods of geologic time and is stable, or unfluctuating, on shorter time scales. This stability may have allowed for the evolution of numerous highly specialized species. Another broad-scale factor that may help account for high species richness is the vast area of the ocean abyss, which offers relatively few barriers to dispersal.[4]

Below the continental shelf, the diversity of the large faunal species increases with depth to a maximum found at an intermediate depth on the continental slope. The diversity then decreases with increasing distance seaward on the abyssal plain. The smaller fauna appear to reach a maximum diversity somewhat deeper. This may be because the food supply for large animals is too scarce in the deepest waters. Some scientists reject the time and area explanations for high deep-sea diversity and prefer biological explanations. One suggestion is that productivity, competition, and predation interact to control diversity.[5] Other theories involve feeding behaviors and how they relate to changing conditions on the sea floor.

Recent samples from the deep-sea floor in the North Atlantic suggest a previously unimagined richness of species.[6] From an area of about 21 square meters (the size of a small room), 898 species were identified, representing more than a hundred families and a dozen phyla. Of that number, 460 were species new to science.

Furthermore, each new sample yielded new species at a steady rate even after 200 samples had been analyzed. Extrapolation from this data suggests that there may be millions of species—giving the deep sea an animal diversity of a magnitude approaching that of the tropical rain forests.

Most species on the deep-sea floor seem to be rare, and the distribution is patchy.[7] This has been attributed to a heterogeneity on a very small scale in a seemingly homogeneous (on a larger scale) environment, a concept first suggested in 1976.[8] As currently refined, the theory is that, given the low productivity, low incidence of disturbance, and large surface area associated with most deep-sea benthic environments, a high species diversity is maintained by small-scale biological dynamics. For example, burrowing by the bottom organisms (and the consequent disturbance of sediments) produces a nonhomogeneous and nonconstant fine-scale topography. Also, the raining down of organic particulate matter that settles to the bottom causes an ever-changing supply and distribution of food. Small particles, such as phytoplankton and zooplankton bodies, tend to accumulate in depressions and burrows, and larger objects, such as fish, whales, and large pieces of debris, cause major changes in the bottom topography and attract congregations of various types of detritus feeders. A blade of eelgrass falling to the deep-sea floor provides a habitat for a unique fauna. Often these special habitats are ephemeral, so species composition may vary temporally as well as spatially.

This biologically induced small-scale variation in the environment has allowed species to specialize and thereby diversify. For example, species may be specialized in the ways they handle their food and in adaptation to physical microhabitats. Over the long periods of time that the abyssal environments have been relatively constant, a very high species diversity has been able to evolve through this process of specialization.

Another variation of the theory suggests that feeding habits of deep benthic animals hold the key to diversity. The dominant lifestyle in the deep sea is described as cropping: the animals feed mainly on detritus raining down from above, but may inadvertently consume any living particles associated with the detritus, including small benthic animals. The larger, mobile scavengers consume and disperse large particles, while small particles such as dead plankton are probably more or less evenly distributed. Thus, the food supply for smaller feeders is fairly consistent. In this setting, the maintenance of high species diversity is attributed to continued biological disturbance rather than to specialization.[9]

The threat to life on the deep-sea floor from pollution has not been given much attention because it has been presumed that at least this environment is out of reach. That, of course, is not true, since the base of the food chain for the deep-sea fauna and much of the food directly consumed by them originates in shallow waters. Deep-sea fish in the Atlantic Ocean have been found to be contaminated with persistent pollutants from land. Also, radioactive wastes, which have proved easier to track in the oceans than organics or heavy metals, provide insight into the magnitude of the problems associated with ocean waste disposal. For example, radioactive materials have been identified in surface waters at Arctic sites where those waters feed the deep ocean, and specific radioactive wastes released from a facility in Great Britain were traced in surface currents all the way to the west coast of Greenland in a matter of a few years.[10]

Pollution in surface waters is carried down to the sea floor through the food chain, in sinking biological detritus, and slowly by way of downwelling waters from the surface in polar regions. Scientists are concerned that toxic pollution and possibly climatic shifts could cause widespread disruptions in the community structure of the deep-sea fauna. The likelihood of extinctions is debated, but deep-sea species have evolved in relatively unchanging physical and chemical environments and are likely to be highly sensitive to toxic pollution.

Hydrothermal Vents

Unusual deep-sea communities have been discovered fairly recently in areas surrounding hydrothermal vents in the sea floor. These communities are part of a unique food chain. Rather than relying on photosynthesis as the primary process fueling the food chain, these communities are dependent on chemosynthesis, made possible by compounds (especially hydrogen sulfide) found in the hot fluid around the vents. Highly productive chemosynthetic bacteria are at the base of the food chain, and they support a highly productive community of invertebrate and fish species unique to this environment. Species diversity at vent sites is low (something over 100 species have been identified to date), but their uniqueness makes them of particular evolutionary interest. These are very unstable environments, subject to fluctuations in the hydrothermal flow and to low concentrations of oxygen and high concentrations of potentially toxic materials such as sulfides, petroleum hydrocarbons, and heavy metals. Most species are common to vent communities in general, but a few seem to be endemic to vents of the individual fault lines along which vents occur.

Submarine Canyons and Other Unstable Areas

Some areas of the deep sea are subject to frequent natural disturbances. These take the form of intense currents, mud slumps, low oxygen, and upwellings. Species diversity in these areas is very much reduced compared with the normal high diversity in more physically stable deep-sea areas. There are fewer species present, and a small number of species dominate the total number.

Deep submarine canyons and trenches (often several thousand meters deep), which cut across continental shelves and slopes, are examples of such unstable areas. The sides of these canyons are exposed rock, so species that attach to hard substrates or hide in crevasses are found here. Often these species are absent from the shelf because of its soft substrates. The canyons are subject to slumping and slides and strong currents, and therefore animals typical of the more stable shelf floor are absent from the canyons. In general, species diversity is low because of the extreme environmental disturbance.

Areas of very strong bottom currents can be found in some locations; for instance, there are deep currents (below 4,000 meters) off the coast of Nova Scotia. In such areas, sediments are continually washed away and the bottom community consists primarily of juvenile stages of very few species.

Open-Ocean Pelagic

Pelagic communities are composed of plankton—plants, animals, bacteria, and viruses that drift with the ocean currents—and nekton, which are animals that swim. Plankton is often divided into phytoplankton and zooplankton. Phytoplankton is normally thought to be composed of assemblages of species of photosynthetic microalgae. However, other types of organisms may also be responsible for photosynthesis in the oceans. Two previously unknown types of photosynthetic organisms have recently been discovered. Both are extremely tiny, primitive, single-celled, and photosynthetic. They are similar to bacteria and live in high concentrations at about 100 meters, where sunlight barely penetrates.[11]

Zooplankton includes animals from single-celled protozoa to large invertebrates, like jellyfish, which move about on ocean currents—not swimming on their own but floating at various depths in the water column. Zooplankton includes animals that are planktonic all their lives as well as larvae of animals that "grow up to be" nekton or benthos. Larvae of benthic species are more likely to be found in coastal pelagic systems than in open-ocean systems. However, it has been pointed out that larvae of both bottom and near-bottom species are carried considerable distances away from shore because of the dynamic exchange between coastal waters and the open ocean, resulting in considerable horizontal fluxes of plankton, including larvae.[12]

It should also be noted that bacteria are proving to be far more abundant in the open ocean than previously thought. Because of the difficulty in identifying species (as previously noted, identification must be done by biochemical analysis of DNA sequences), it is impossible to say how many different species are involved and how they are distributed. However, using the few well-known and well-studied species as indicators, their distribution patterns appear to follow physical barriers in the ocean (currents and vertical gradients) similar to patterns followed by other types of organisms.

The life in open-ocean pelagic communities at all depths is not only highly varied; it includes mysterious and fantastic forms that are far beyond our daily experience on land. For example, phosphorescence (biochemical light production), which is a frequently observed phenomenon in the sea, is relatively uncommon on land. We are familiar with lightning bugs, but in the sea phosphorescence is found in a myriad of species, from phytoplankton to deep-sea fish that glow with elaborate patterns of lights. There are clearly many tales to tell about biological form and function, food web dynamics and interrelations among species, in open-ocean pelagic communities; but most of these stories are as yet unknown to science.

The open-ocean pelagic systems are physically defined by large, stable circulation patterns, such as large-scale permanent gyres in the Atlantic and Pacific oceans. The gyres circumscribe horizontal areas, but the systems are also vertically stratified, with characteristic plankton and nekton diversities at several depth intervals defined by physical gradients. Species diversity within these ecosystems has been studied for individual groups of animals, especially zooplankton, but not for the entire community. As Ramon Margalef has noted, "Total diversity is almost mythical," but diversity of the ecosystem "is reflected with little distortion at several levels."[13] To be good indicators of total diversity, the level, or taxonomic group, must be distributed across the whole ecosystem. Therefore, phytoplankton and zooplankton are good indicators. When considering the biology of pelagic ecosystems, it is important to look at gradients, because the composition and properties of plankton and nekton within neighboring regions are interdependent, the result of the dynamic exchange that is characteristic of fluid environments.[14] It has been noted by several observers that gradients in plankton species diversity are related to gradients in productivity—the relationship generally being inverse. This certainly holds true for what we know about productivity in the open ocean, but there is still much to be learned and therefore it can only be a working hypothesis. There is some evidence that latitudinal gradients and distance-from-shore gradients hold true for pelagic communities and may correspond to gradients in the physical environment and its variability. For example, the diversity gradient in krill species (tiny crustaceans) increases with decreasing latitude.

Sargasso Sea

Major pelagic systems that have been studied with respect to zooplankton species diversity include the Sargasso Sea, which is defined by the major central gyre of the North Atlantic. The species diversity of the Sargasso Sea is characteristically high compared to inshore pelagic communities.[15] This area is typical of tropical Atlantic waters, where plankton density is low but the number of species is high, consistent with the gradient of increasing diversity with distance from shore. Oligotrophic waters (low nutrients and low primary production) of the Sargasso Sea in the North Atlantic have a greater species diversity than do the waters of the continental slope; these in turn have a higher diversity than waters overlying the continental shelf.[16] The higher diversity of the Sargasso Sea is thought to be a result of greater environmental stability in the deep waters relative to coastal waters, where there exist environmental instability and seasonal fluctuations. Recent genetic studies of bacterioplankton in the Sargasso Sea revealed an interesting diversity of microbes with a relatively broad spread of genetic variation.[17]

Included in the rich species diversity of the Sargasso Sea is a unique community of floating seaweed and associated fauna. It is the only open-ocean ecosystem that includes macroalgae—in this case, *Sargassum*, a genus commonly found attached to rocks along continental shores of the North Atlantic. The *Sargassum* occurs as floating rafts, and it is clear that these are not merely drifting remnants of seaweed torn up from some shore. Living among these rafts of *Sargassum* is a community of animals that have clearly evolved to live in this specific environment, for they are almost without exception the same color as the *Sargassum* and many have decorative proliferations that make them resemble the seaweed itself—natural camouflage at its finest!

Systems of the Pacific Ocean

In the open-ocean environment of the Pacific, patterns of most species assemblages strongly resemble those of the major circulation systems of the Pacific: the North Pacific Subarctic cyclonic gyre; the Central North Pacific anticyclonic gyre; and the Eastern Tropical Pacific. There is agreement as to the shape of the patterns of species distributions within most higher taxa; e.g., copepods, euphausids, pteropods, thaliaceans, and even bacteria exhibit patterns of distribution similar to the current patterns.

The North Pacific Subarctic cyclonic gyre is characterized by massive upwelling of deep water and very strong seasonal cycles in both hydrography and biology. The area is rich in production but poor in species diversity—at least for plankton. In contrast, the Central North Pacific anticyclonic gyre is characterized by sinking waters in the interior, due to an inverted halocline, and by higher temperatures in the upper stratum. It is a relatively homogeneous, nonseasonal system of low production but high diversity—"the most species-rich province in the Pacific."[18]

The Eastern Tropical Pacific Ocean, while not a gyre, is a well-defined area near the equator where nutrient-rich waters converge. It is characterized by upwelling subunits—including the Costa Rica dome, among the most productive of tropical waters—and the upwellings in the Gulfs of Panama and Tehuantepec. The nutrient-rich Peru Current flows into this system from the southeast. The sources of these rich waters are variable, and they have different temporal scales. Thus, the system is heterogeneous but productive. Diversity has not been well studied, but seems to be only moderately rich, possibly reflecting a common relationship between high productivity and reduced diversity.

The central gyre of the North Pacific is considered to be the closest approximation to a semiclosed ecosystem in the open ocean; it is defined by gyrelike circulation, large-scale and persistent features of mixing, endemic physical processes, and local climatic pattern. A study of copepods, crustaceans that have the largest number of species and number of individuals and the greatest biomass of all the zooplankton, demonstrated that the species exhibit patterns of vertical zonation within the gyre.[19]

This zonation effectively divides the pelagic zone into separate ecosystems, each with its own characteristic species assemblages and each dominated by a particular trophic level. In other words, the surface zone is dominated by grazers, which use phytoplankton as a food source; beneath the surface zone are successive levels of zooplankton feeders; and near the bottom detritus feeders become dominant. Within each zone, species compete for the same general food source, and competitive exclusion of the less effective species may be regulated by selective predators (fish feeding on the more successful species of zooplankton, for instance). On the other hand, some oceanographers believe that physical phenomena control species diversity in pelagic communities, with competition and predation playing a secondary role.

Species assemblages of fish seem to follow much the same pattern as those of zooplankton, confirming the usefulness of one group as an indicator of species diversity in general. In both the Central North Pacific gyre and the South Pacific gyre (a large central gyre in the Southern Hemisphere), fish assemblages have been found to be rich in species, but there are twice as many species in the Central North Pacific. These two systems are self-contained, self-regulated, and semiclosed ecosystems with a high degree of biological regulation.[20]

Two stable assemblages of phytoplankton species have been identified in near-surface waters of the Central North Pacific gyre ecosystem. One group is associated with a surface zone where nutrients are limiting, and the other group in a lower zone where nutrients are more plentiful but light is limiting. The maintenance of diversity within the zones has been attributed to predation in the upper community and competition in the lower community.[21]

There have been some interesting observations indicating that preconceived ideas about the sparseness of phytoplankton and its low productivity in these waters may be incorrect or productivity may be changing. Measurements of high concentrations of oxygen isolated in a water mass in the Central North Pacific gyre indicate that phytoplankton productivity may be several times higher than earlier thought. Also, significant increases in total chlorophyll during the summer were observed between 1968 and 1987. These changes have been attributed to physical factors: an increase in winter winds and a decrease in sea-surface temperature contribute to increased vertical mixing, providing more nutrients to the surface waters from below.

Excursions into deep waters by manned submersibles have provided reports of dense concentrations of bacteria and organic matter at various depths. The importance of this biological material to the pelagic and deep benthic food chains cannot be overemphasized.

Polar Seas

The Arctic Ocean and the seas surrounding Antarctica are special pelagic ecosystems that include a solid phase—ice—in the physical environment. These waters in general are rich in nutrients, obviously cold, and the ice interferes with light penetration into the waters below. The primary producers (microalgae), which require light, have adapted to the situation by living in the ice itself or adhering to the underside of the ice. So do bacteria, protozoa, and some of the zooplankton (e.g., krill, copepods). These organisms form the base of a food chain that ultimately supports numbers of fish, seabirds, and sea mammals. The ice also functions as a platform from which seals and polar bears search for food, and upon which seals breed.

Diversity in the Arctic and Antarctic is increased by a system of leads and polynyas; they are created in areas where ice forms but is driven away by winds, or where rising water from the deep causes concentrations of warm water where no ice forms. These are productive areas, rich in nutrients and sunlight, and provide hospitable environments with plentiful food for seabirds and mammals. Many of these open areas occur in the same location year after year, and are therefore reliable places for the birds and mammals to congregate. In addition, the heterogeneity of the ice itself may provide for increased diversity of microscopic species living within that environment, though that diversity is not thought to be extraordinarily high. As already mentioned, the benthic Arctic environment does not support a high diversity of species.

Surprisingly, a major threat to Antarctic seas is pollution from scientific research facilities and from a new tourist industry.[22] The scientists have been alerted to the problems they are creating, and the National Science Foundation intends to make a concerted effort to prevent future pollution and to clean up debris, garbage, and toxic waste from past activities where possible.[23] Another looming threat is the exploration of minerals by several nations. Currently, there is an international treaty being negotiated that would regulate such exploration, but environmentalists and some nations would prefer to prohibit mineral exploration altogether and make Antarctica an internationally protected wilderness area.

The Arctic Ocean currently suffers from overfishing and from the effects of oil exploration and transportation. Habitats are disrupted, and hydrocarbons pollute the waters. The 1989 *Exxon Valdez* oil spill in Prince William Sound, Alaska, drove home the vulnerability of arctic and subarctic ecosystems to oil exploration on arctic shores and in arctic waters, and brought into question the wisdom of such activity. Proposals to exploit oil resources in the Arctic National Wildlife Refuge and the Chukchi Sea are ecologically foolhardy, if not immoral.

Another threat, perhaps even greater, may lie ahead: global warming due to increased carbon dioxide in the atmosphere due to industrialization. The greatest changes in the physical ocean environment are predicted for the Arctic. Temperatures would rise and the ice would melt. Melting of the ice would probably be more significant to the biota than the rise in temperature. Although it is likely that warmer water species would invade Arctic habitats, confining Arctic species farther north, and although the outcome of interspecific competition is not predictable, the destruction of a very important habitat—the ice itself—could threaten those species not able to adapt successfully to life in liquid water. The full impact of runoff from a thawing tundra is also difficult to predict.

There might also be physical changes in the Arctic seas that would have repercussions through all the oceans by altering the patterns of circulation, food chains (including valuable fisheries), and climate in other parts of the world. The melting ice would contribute significantly to sea-level rise and thereby compound the negative effects of that problem.

Chapter Five ▌

Conservation of Marine Biological Diversity

HUMAN tinkering with natural ecosystems inevitably interferes with their dynamic homeostasis and, more often than not, results in the loss of biological diversity. This is true of well-planned protective management practices as well as intentional exploitation. As Ramon Margalef has stated in his *Perspectives on Ecological Theory:* "Any human intervention in nature, even presupposing good intentions, can rarely be reconciled with the idea of strict conservation. . . . Genuine conservation forbids any interference."[1] The conservation methods practiced by environmental managers and regulators are, at best, a compromise between naturally evolving ecosystems and the overly exploitive human species.

Since models for protecting biological diversity have been developed on land—where it is far easier to assess the loss of species—it is important to understand the differences in the two environments in order to judge the potential effectiveness of transferring terrestrial conservation techniques to marine environments. As Rodney Salm (1984) so aptly put it:

> It is the special burden of marine conservationists that people cannot easily see what happens under water. The sea remains inscrutable, mysterious to most of us. On land we see the effects of our activities and we are constantly reminded of the need for action, but we see only the surface of the sea. Not only are we less aware of our impact on submerged life, but it is also more difficult to investigate.[2]

Comparisons of biological diversity on land and in the ocean will necessarily involve real differences between marine and terrestrial biotas as well as differences in how the two are studied and what is known about them. For example, species tend to be the focus of conservation efforts on land, but that focus is not very helpful in the sea. Terrestrial species are more familiar to us and their distributions more easily determined. They are more readily mapped, cataloged, and counted, and it is easier to determine for terrestrial species whether populations are so limited that a species should be classified endangered.

Efforts directed specifically at protecting biological diversity in the marine environment have been limited, primarily because diversity in the oceans is not widely considered to be threatened. In terrestrial environments, the protection of biological diversity has centered around three approaches: establishing protected habitats, such as national parks; preventing the overexploitation of threatened and endangered species; and establishing living gene banks, such as zoos, botanical gardens, and seed banks. While there is value in the first two approaches for the sea, the fluid nature of that environment dictates a different focus. Methods of protecting ocean diversity must be developed around the character of the marine environment and should not be a simple transfer of terrestrial methodology into the sea.

It is difficult to protect marine biological diversity by drawing lines around habitats or particular species, because the fluid medium will not honor those boundaries. Toxic pollution will be carried into the "safe" zone, larvae and motile species will be carried out. Furthermore, restricting activities such as shipping and oil ex-

73

ploration is proving difficult in marine protected areas, and the issue of fishing restrictions remains a problem. The question of just what will be allowed and what will be disallowed in marine reserves has yet to be resolved.

However, no matter how strict the regulations are within marine protected areas, the marine environment and its biological diversity will not be adequately protected without appropriate regulation of development in the coastal zone—including the multitude of interconnected uses of (and consequent impacts on) land, air, and ocean by human populations. What is needed is integrated coastal management on a regional scale and restriction or regulation of ocean uses (e.g., discharge of wastes, fishing) consistent on a global scale. In addition to direct regulation of coastal development, regulating water and sediment quality, and phasing out toxic discharges into marine waters may be effective in protecting biological diversity, even though that is not usually the express goal of such measures.

International agreements and national laws are needed to drive the solutions, but implementation must be on the local level. Consequently, coastal communities must be motivated to participate in effective coastal zone management, and they must have access to the appropriate scientific guidance and environmental monitoring to ensure the effectiveness of their programs. Here is where economics comes into play, and the use of economic incentives is now seen as a critical component of programs to conserve biological diversity. Also on the horizon is a promising new approach to economic valuation of the environment: the assignment of dollar values to noncommercial resources such as biological diversity. Another economic principle that is gaining favor with the public is the notion that the "polluter must pay" for environmental damage through a system of fees for permitted discharges and stiff fines for exceeding permitted discharges.

Abatement of Marine Pollution

Since pollution is a leading threat to marine ecosystems and their biological diversity, the policies of human societies and governments with respect to water pollution are critical to the conservation of marine diversity. As a result of ineffective policy, many coastal marine ecosystems have been degraded by pollution from runoff and municipal and industrial outfalls. However, there is reason to believe that many coastal areas could recover from those effects if a strong antipollution policy were adopted on regional and global scales. The deep ocean is in an even better position to benefit from such a policy, since signs of pollution in those waters are just beginning to appear and there is no evidence of impoverishment of those biological communities. However, it is important to note once again that the absence of evidence should not be interpreted as evidence of absence of effects. Many of these open-ocean ecosystems have not been studied well enough for us to be assured that they are not damaged by pollution.

There is currently enough interest and discussion of marine pollution policy at national and international levels that it seems likely that new policies with respect to marine pollution will be established in the near future. However, it is not at all clear what those policies will be. Three widely differing approaches are advocated: (a) the intentional use of the vast oceans for the disposal of wastes from land as an answer to the burgeoning waste problems of human societies, particularly the more "developed" societies; (b) the regulation of discharges to maintain an "acceptable" level of pollution which might not adversely affect living marine resources; and (c) the phasing out of pollution, which would involve dramatic changes in life-styles and the economic goals of people in both "developing" and "developed" nations.[3]

Where policies have been initiated, the middle ground has been chosen. However, it is becoming more and more apparent that pollution and living marine resources are incompatible, and it is impossible to strike a balance that will allow the use of the oceans for waste disposal and to maintain diverse marine ecosystems. The constant struggle and compromise between factions wanting stricter and those wanting more lenient pollution controls invariably end up with politically dictated levels of allowable pollution. This process cannot be expected to be in harmony with the actual tolerance levels of biological communities; so it is not surprising that in countries where such a policy has been implemented—the United

States, for example—coastal ecosystems are becoming degraded despite regulations that limit polluting activities.

The continued decline of marine systems suggests the need for tighter pollution controls and greater economic burdens to be placed upon polluters. Some environmentalists realize that the only effective solution is to eliminate water pollution at the source, a goal that would require more than just stricter regulations. It would require a whole new approach to industrial development and to activities that result in significant runoff, such as farming. New production methods that do not produce toxic waste would be required, as would the redesign of many products to eliminate sources of pollution. Agricultural practices would have to abandon the heavy reliance upon chemical fertilizers and pesticides. Such a dramatic policy change is beginning to gain favor in many international circles as it is recognized that the ocean is a vital living system that may hold the only hope for maintenance of the biosphere in the face of the escalating destruction of terrestrial ecosystems.

Antithetical to such a precautionary approach to ocean pollution policy is the suggestion that the ocean should be used to the fullest as a waste bin for human societies. The advocates of this policy— often including ocean scientists who are not biologists—see the ocean as a gigantic pool of liquid available to dilute away the biologically hazardous by-products of industries as they are now designed. This approach does not factor in the consequences to the living component of the ocean and the repercussions these impacts would have on terrestrial life.

As ocean policy alternatives are debated in national and international arenas, some governments have recognized the need to follow the middle course, at least, and take some action to control pollution in coastal environments. To date, the principal method of reducing marine pollution has been through national legislation and international agreements or conventions. The laws begin the process of controlling or eliminating pollution, but effective implementation and enforcement are needed to achieve the goals stated in the laws.

Laws are implemented through specific regulations, and if those regulations are not strong enough, the pollution will continue to be a problem for living marine communities. Even if the regulations are adequate, the poor enforcement of those regulations may be a weak link to achieving adequate protection. Finally, the effectiveness of the regulations must be measured through environmental monitoring. To protect biological diversity, the diversity itself must be monitored. When monitoring reveals that protective measures have been inadequate to prevent damage to living marine systems, it will be necessary to strengthen the laws, regulations, and enforcement.

There are basically two approaches to regulating pollution: by establishing water quality objectives and by directly regulating discharges. Both are needed for effective pollution control. Water quality standards developed for both freshwater and marine coastal waters are helpful in reducing ocean pollution, since much of the pollution enters the ocean via rivers. These standards are numerical or descriptive limits on the concentration of certain pollutants that will be allowed in a body of water. If water quality standards are exceeded, sources of pollution must be assessed and significantly reduced.

Sediment quality standards for coastal waters are an often ignored but very important component of pollution abatement, because sediments often act as a sink for pollutants entering coastal waters and subsequently a source of pollutants entering the marine food chain. Sediment quality standards should directly apply to dredging and dredge spoil disposal activities, the cleanup of contaminated sediments, and identifiable activities that threaten to continue to contaminate sediments. Air quality standards are also important to the reduction of marine pollution, since atmospheric deposition is a major source of some pollutants in the ocean.

The direct regulation of discharges should apply to sources of polluted runoff (also known as nonpoint sources) as well as industrial pipeline discharges, sewage treatment plant discharges, and sewage plant overflows (known as combined sewer overflows, or CSOs). In practice, however, nonpoint sources—which include agricultural and urban runoff—are much more difficult to regulate, and it will be a long time before laws, regulations, and enforcement are effective in preventing nonpoint source pollution.

In contrast, direct discharges are relatively easy to identify and regulate, but implementation and enforcement of regulations are often inadequate. A policy of leniency toward industrial dischargers is common, even when, by law, direct dischargers of waste water—e.g., from sewage treatment plants and industries—must treat the waste prior to discharge to a level consistent with the best available pollution-control technology. Uniform emission standards may be established by category of discharge, such that all dischargers in that category must meet this technology-based standard.

The regulation of direct discharges into marine waters from vessels has been somewhat more effective on both national and international levels. For example, global conventions have been adopted to control the discharge of oil, chemicals, sewage, and garbage, including plastics from vessels. While enforcement remains a problem, these international agreements are believed to have curtailed the discharge of pollution from ships. While the at-sea disposal of "hazardous" wastes is regulated, and in some instances prohibited by international law, land-based sources are, by and large, regulated on a nation-by-nation basis. International action to control land-based sources of marine pollution has been taken in only a few regions of the world. These include the North Sea (Convention for the Prevention of Marine Pollution from Land-Based Sources, 1974—Paris Convention) and the Baltic Sea (Convention on the Protection of the Marine Environment in the Baltic Sea Area, 1974—Helsinki Convention). Some of UNEP's Regional Seas Conventions have also been supplemented by land-based sources protocols. These include the Mediterranean (Athens Protocol to the Barcelona Convention) and the Southeast Pacific (Quito Protocol to the Lima Convention). Since about 70–80 percent of marine pollution comes from land-based sources, that is clearly an important issue to be addressed by a comprehensive international convention.

Integrated Resource Management

The observation has been made that the world is heading toward a future of total environmental management by humans.[4] In this view, first the landscapes and then the seascapes will be controlled. In other words, humans will determine what species grow where and for what purpose, and probably which species go extinct and which are preserved. The outcome of this process would clearly depend on how knowledgeable the managers are about the needs for maintaining ecosystems and the biosphere. Even if it is conceivable that ecosystems can be totally managed so that the biogeochemical cycles of the living planet are kept in balance, it is hard to imagine humans attaining the knowledge necessary to accomplish that kind of management.

While the wisdom of trying to manage the earth on a global scale is highly questionable, the management of some smaller systems is inevitable—and, sometimes, the only means of protecting an ecosystem from worse fates. As human populations dominate the landscape, the management of activities that have environmental impacts has been applied as a means of directing the effects to the least harmful locations and reducing the magnitude of impacts.

The coastal areas of the world are under intense pressure from burgeoning human populations and rapid urbanization and industrialization, and it is becoming clearer that piecemeal regulations and management practices will not save ecosystems of the coastal zone, where so many types of human activities have cascading effects on different segments of the environment. For example, extensive inland agriculture can contaminate groundwater and the rivers that carry this contamination into estuaries, open coastal waters, and on out to sea. Similarly, industrial air pollution can lead to significant pollution of coastal waters through air-sea interactions.

Consequently, what is needed are enforceable regional coastal zone management plans that take into consideration the combined effects of all types of land use and coastal development and provide guidelines and regulations to control this development. This type of management approach is known as integrated resource management, and it is being applied, at least in theory, to some coastal regions of the developing world. The United States has yet to adopt a truly integrated form of coastal management, despite the fact that the concept of coastal zone management originated in the United States.

A good management plan will accommodate modifications in the management practices to respond to the results of monitoring. This process, known as adaptive management, is as important to integrated coastal resource management as it is to the management of individual activities.

Management programs for offshore marine systems are also being developed on a regional scale. Known as large marine ecosystems (LMEs) or regional seas, these management units are seen as biologically the most meaningful scale for managing living marine resources. Up to now, resource managers have focused exclusively upon target species with respect to fishing pressures. The LME approach offers an ideal opportunity to develop a broader management outlook that includes the biological diversity of the system. Since resource species are integral pieces of whole ecosystems, ecological processes, including interactions with other species, will affect the population dynamics of the individual resource species. Consequently, it is important for managers to pay attention to the system as a whole.

The maintenance of biological diversity is an important component of regional management plans in the coastal zone and in large marine ecosystems offshore. The designation of protected areas within the planning region is one appropriate tool for conservation, but the mitigation of impacts on biological diversity throughout the region is equally important.

Economic Incentives

Beyond the laws, international agreements, and management plans, there are a number of effective economic incentives that can be applied at local, national, and international levels to promote conservation of biological resources and biological diversity. An excellent book on the relationship between economics and biological diversity, with guidelines for implementing economic incentives, has recently been published by the International Union for the Conservation for Nature and Natural Resources (now known as the World Conservation Union)[5] and should be consulted for an in-depth discussion of the topic. A promising new field of "ecological economics," which combines the disciplines of ecology and economics, has emerged. The new academic journal *Ecological Economics* states that such integration "is necessary because conceptual and professional isolation have led to economic and environmental policies which are mutually destructive rather than reinforcing in the long term."[6]

In the developed world, where industrial pollution and polluting by-products of energy consumption are a major threat to the environment—especially the aquatic environment—a system in which polluters pay to pollute can be an effective economic disincentive. One approach is to charge polluters for cleanup after the damage is done, but this may be less effective than charging (or taxing) a polluting activity from the onset. If the penalty is great enough, this approach should make it more economical for the potential polluter to reduce pollution at the source and thereby avoid environmental damage in the first place.

One of the great challenges of the future is to convince developing nations that following in the footsteps of the wealthy developed countries is not an advisable goal. The disastrous effects on biological resources are predictable and the planet cannot support that kind of development (and life-style) in the face of exploding human populations. Development cannot occur with the abandon that industrialized nations practiced in the past. The developed nations can begin to meet this challenge by mending their ways and setting an example based on frugal energy consumption, recycling, and sustainable use of resources.

Once they have demonstrated a willingness to restrict their own behavior, the wealthy nations can more effectively apply economic incentives to developing nations—for example, the "debt for nature" swap (forgiving foreign debts if the debtor will agree to conserve specified natural resources), or establishing strict environmental impact restrictions associated with the lending policies of international development banks.

It is also incumbent upon the wealthy lending countries to transfer to borrowing countries appropriate management and technology techniques for mitigating and avoiding negative environmental impacts of development. The best ways of implementing such management and technology will have to be region-specific and management will be most effective if it comes from within the developing nations themselves. In short, both developed and developing nations must become more responsible in directing future development toward sustainable uses of natural resources.

To protect biological diversity not directly linked to commercially valuable products, a new approach to economics attempts to assign dollar values to biological resources that are not directly

81

consumed or marketed. Such an approach may facilitate a new direction to development by placing a value on species and ecosystems for their nonconsumptive benefits, including tourism, science and education, climate regulation, maintenance of water and mineral cycles, and the maintenance of a healthy environment for humans, as well as the potential value of future uses and the value of fulfilling ethical obligations to nonhuman life.

Protected Areas

Perhaps the most popular conscience-easing way of trying to protect species diversity is by setting aside reserves within which the flora and fauna are protected. This has been done extensively in the terrestrial environment of the United States (and also in other countries) with the establishment of protected areas for designated uses and for multiple uses. The first category—protected areas for designated uses—is by far the most effective in protecting biological diversity and includes national and international parks, wilderness areas, and wildlife refuges. Some of these protected areas include coastal lands with marine intertidal communities, such as marshes. Far less restrictive are the multiple-use management areas, where development is restricted but exploitation of natural resources is allowed or even promoted. Often these areas claim to allow resource exploitation only up to the sustainable level, but it is arguable whether that level is adequately assessed, as, for example, in the national forests of the United States and Canada.

The terrestrial multiple-use model is being followed in the marine environment despite the significant differences in the nature and perceptibility of threats to biological diversity in the two environments. Parks with very strict prohibitions are uncommon in the underwater world. More often, marine protected areas are identified as rare and valuable habitats but do not carry explicit prohibitions against multiple use. Both nationally and internationally, many marine sanctuaries are seen as multiple-use management areas rather than as biological preserves.

Although protection of selected marine habitats should be encouraged and expanded, it must also be recognized that there are limitations to the effectiveness of this strategy in preserving marine biological diversity. The two major weaknesses of this approach are the fluid nature of the environment, which prevents the exclusion of pollution from protected areas, and the biological interdependency of neighboring ecosystems. Animals targeted for protection commonly migrate in and out of a sanctuary, thus reducing the effectiveness of the protection. Migrating birds face the same problem in terrestrial environments—while they may be protected in one part of their annual range, they may be threatened in another part. Consequently, marine species are not going to be saved solely by establishing protected areas. Nevertheless, a global system of marine and coastal protected areas could go a long way toward protecting particular habitats in the world's oceans. These areas, however, should be integrated within a broader national marine conservation strategy encompassing the other tools described in this section.

Regulation of Living Marine Resources

In the marine environment, certain species are exploited for their commercial value as food, as sources of valuable products (e.g., oils, furs, medicines), and as souvenirs (e.g., seashells, corals). Overexploitation of such species can be controlled by managing the harvest, supplementing the harvest with cultivated stock, and prohibiting the harvest and trade of endangered and depleted species.

National and international regulation of fisheries helps conserve valuable fisheries species but usually does not take into account the nontarget fish species harvested or adequately deal with the incidental catches of mammals (especially porpoises), turtles, and seabirds. Furthermore, the depletion of many fisheries indicates that some fisheries management efforts are not stringent enough to be effective. The more complicated problem of identifying and protecting species whose populations may be depleted because of their dependence upon the harvested fisheries species is also not addressed by any management programs.

Fisheries management programs need to focus on fishing technology as well as on catch limits. Certain modern technologies are proving to be devastating not only to populations of target species but to a vast diversity of marine life. Examples of methods that are contributing to the depletion of living marine resources include (a) fishing the deep ocean with huge drift nets, (b) operating large vessels able to process huge catches at sea, (c) using aerial spotters and acoustic fish-finders to locate schools of target fish, and (d) employing more and more effective fishing equipment without restrictions on size or location of catch. Overexploitation due to overblown technology will have to be controlled at the international as well as at the national level, since much of the activity takes place on the high seas.

The unregulated harvest of whales for their oils in the 1800s is, of course, an infamous case of overexploitation, one that drove many species to the brink of extinction. A similarly devastating harvest of marine mammals for furs (seals, sea lions, otters) imperiled many more species. The world is still struggling with the severe depletion of many marine mammal species and the problem of protecting them from further exploitation. International bans or limits on whaling have been effective to the extent that major whaling nations participate, but not all do. Bans or limits on hunting and trade of marine fur mammals have helped slow their march toward extinction. Official recognition of "endangered and threatened species" by many nations has reduced but not eliminated the harvest and trade of these species.

The identification of endangered species lags far behind that on land, in part because so many marine species are unknown and/or unassessed as to their populations. It is also widely believed that, except for marine mammals and turtles and some marine birds, endangerment is not a problem in the oceans. There is no sound data to support this assumption, however, so a greater effort is needed to ascertain the status of marine species. In fact, the recent realization that coral reefs are in trouble has led to the addition of several species of corals to the list of species regulated by international trade agreements.

Efforts to reestablish populations of some highly visible marine species have met with limited success. For example, the sea otter on the west coast of the United States is extending its badly whittled distribution through controversial relocation programs; also, there is a natural expansion of populations in formerly inhabited areas now that the species is protected. The Atlantic salmon has been returned to some of its natural habitats along the northeastern coast of the United States through restocking programs. Generally, however, these efforts have been limited to endangered mammals or economically valuable species.

Living Gene Banks

Living gene banks for marine species are less common than for terrestrial species. Aquariums are not nearly as numerous as zoos, and they have no demonstrated role in preserving genetic stocks of the species they display. There has been a claim that aquariums, and other institutions that exhibit marine mammals, have the multi-purpose function of providing entertainment and education to the public as well as serving as a refuge for genetic stocks that can be used to replenish wild populations. However, the rate of success of marine mammal breeding in captivity is not high and most aquariums do not have the facilities for breeding. Before approval is granted for captivity of endangered marine species, there should be better demonstration of the effectiveness of captivity in protecting the species.

Hatcheries and cultivation projects for economically valuable marine fish and shellfish are of some value as gene banks. Hatcheries, for instance, continually replenish wild populations of fish or shellfish in an attempt to keep populations of certain commercial species at a healthy size. Since the source of genetic material is usually wild stock, hatcheries effectively help preserve genetic variation. Mariculture, on the other hand, may or may not use wild populations as breeding stock. If not, the genetic diversity of the cultured populations is reduced. Selective breeding has not been practiced much in mariculture, and the kind of genetic variety available in seed banks of terrestrial species for agricultural purposes has no counterpart in the sea.

Is Marine Biological Diversity Currently Protected?

Although biological diversity has not been the focus of most marine environmental protection initiatives, either nationally or globally, there are existing laws, conventions, and programs that indirectly afford some protection to marine diversity. The next two chapters of this book briefly examine U.S. provisions and programs as well as the international agreements and programs pertinent to the conservation of marine biological diversity. Unfortunately, diversity is rarely assessed in connection with these initiatives, so it is difficult to measure their success as conservation tools.

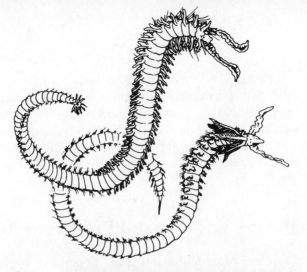

Chapter Six

National Programs and Laws

UNITED STATES efforts to protect biological diversity in the marine environment have been sporadic at best. It would appear that there are numerous laws and programs established to protect marine biological diversity; yet clearly this is not the case. Due to inadequate understanding of the importance of maintaining diversity in marine systems, laws and policies have been enacted on an ad hoc basis to deal with particular threats. The United States does not currently have a comprehensive marine environmental protection strategy. Conflicting demands placed on the marine environment, coupled with conflicting legal and administrative mandates, have led to compromise and ultimately the destruction of valuable marine and coastal habitats.

While the following laws and programs are a significant commitment on the part of the United States to protect the marine and coastal environment, a more comprehensive effort is necessary. Laws and regulations must do more than buy time by slowing down the degradation of the marine environment; they must restore damaged ecosystems and prevent the defiling of healthy ones.

Current U.S. programs and laws can roughly be divided into three groups: marine and coastal protected areas, prevention of the overexploitation of species, and protection of water and sediment quality. This list should not, however, lead one to the conclusion that all these areas are adequately covered by legislation and existing programs.

Marine and Coastal Protected Areas

National Marine Sanctuaries Program

In 1972, Congress, recognizing that certain areas of the marine environment possess "conservation, recreational, ecological, historical, research, educational, or aesthetic qualities which give them special national significance,"[1] established the National Marine Sanctuaries Program. This program is the only federal program specifically designed to protect biological diversity in the ocean. The purposes of the program are:

- to identify areas of the marine environment of special national significance due to their resource or human-use values
- to provide authority for comprehensive and coordinated conservation and management of these marine areas that will complement existing regulatory authorities
- to support, promote, and coordinate scientific research on, and monitoring of, the resources of these marine areas
- to enhance public awareness, understanding, appreciation, and wise use of the marine environment
- to facilitate, to the extent compatible with the primary objective of resource protection, all public and private uses of the resources of these marine areas not prohibited pursuant to other authorities.

Under the Marine Sanctuaries Act (Title III of the Marine Protection, Research, and Sanctuaries Act), the Secretary of Commerce is given authority, after consultation with Congress and with other federal and state agencies, citizen groups, and the general public, to designate marine sanctuaries in any area of the marine or Great Lakes environment within the jurisdiction of the United States (i.e., within the U.S. 200-mile exclusive economic zone). In making the determination, the Secretary, acting through the National Oceanic and Atmospheric Administration (NOAA), must consider:

- the area's natural resource and ecological qualities, including its contribution to biological productivity, maintenance of ecosystem structure, maintenance of ecologically or commercially important or threatened species or species assemblages, and the biogeographic representation of the site
- the area's historical, cultural, archaeological, or paleontological significance
- the present and potential uses of the area that depend on maintenance of the area's resources, including commercial and recreational fishing, subsistence uses, other commercial and recreational activities, and research and education
- the present and potential activities that may adversely affect the area's resources
- the existing state and federal regulatory and management authorities applicable to the area and the adequacy of those authorities
- the manageability of the area, including factors such as its size, its ability to be identified as a discrete ecological unit with definable boundaries, its accessibility, and its suitability for monitoring and enforcement activities
- the public benefits to be derived from sanctuary status, with emphasis on the benefits of long-term protection of nationally significant resources, vital habitats, and resources that generate tourism
- the negative impacts produced by management restrictions on income-generating activities such as living and nonliving resources development
- the socioeconomic effects of sanctuary designation.

Although the sanctuary program is not a strict "wilderness" program in the traditional sense and calls for "multiple use," the overriding consideration is the protection of the natural resource values of the particular area. The law does not specifically prohibit any activity within a marine sanctuary; rather, NOAA is given broad authority to regulate any activities that are not compatible with resource protection. Activities prohibited or stringently regulated in the various marine sanctuaries to a greater degree than called for in other statutes include oil and gas development, harvesting of living marine resources, dumping or discharging of any wastes, vessel traffic, alteration or construction on the seabed, and other activities that may harm sanctuary resources.

The management of marine sanctuaries represents a significant change from single-purpose marine legislation—which regulates, for example, dumping of wastes or harvest of marine fisheries. The Marine Sanctuaries Act and implementing regulations take a balanced look (in theory, if not in practice) at the uses of the marine environment, with a consideration of that environment as containing inherent natural values worthy of protection and restoration.[2] As such, the entire area, or ecosystem, within a sanctuary is managed as a whole.

The national marine sanctuaries (NMS) designated by 1990 include the following.

- The Channel Islands NMS (desig. 1980), located off the shore of southern California, is the largest of the sanctuaries. It includes the Channel Islands and Santa Barbara Island, and protects a 1,252-square-mile habitat for marine mammals and seabirds.

- Cordell Bank NMS (desig. 1989) is a seamount located at the edge of the continental shelf about 20 miles west of Point Reyes, California. It rises to within 115 feet of the ocean surface. The sanctuary encompasses about 100 square miles and contains a distinctive benthic community and a rich feeding ground for fish, sea mammals, and seabirds. The sanctuary is the first to exclude oil and gas exploration within its boundaries, a restriction guaranteed by an unprecedented law, specific to Cordell Bank, passed by Congress in 1989.

- Fagatele Bay NMS (desig. 1985) is located in American Samoa and is comprised of 160 acres of a typical terraced coral reef ecosystem associated with high volcanic islands of the warm water areas of the Pacific Ocean.

- Florida Keys NMS (desig. 1990) covers over 2,600 square nautical miles of ocean waters. This sanctuary encompasses the former Key Largo and Looe Key sanctuaries and extends over a vast ocean area from Biscayne National Park to the Dry Tortugas on both the Straits of Florida and Florida Bay sides of the Keys. The newly designated sanctuary harbors fragile coral reef communities. Legislation prohibited oil and gas development and imposed restrictions on vessel movement within the boundaries of the sanctuary.

- Flower Garden Banks NMS (desig. 1990) is located in the Gulf of Mexico about 120 miles from the shore of the Texas-Louisiana border. The sanctuary includes two banks raised by intrusions of salt plugs from below the sea floor, which are covered by luxuriant coral reef and algal reef communities. Also included in the area is an unusual and distinctive ocean brine seep community associated with high salinity and with sulfide-rich waters (which emanate from the surface of one of the salt plugs).

- Gray's Reef NMS (desig. 1981) covers a 17-square-mile area off the coast of Georgia and is situated in a transition zone between the warm Gulf Stream and more temperate coastal waters. It is composed of one of the largest "live bottom" reefs on the South Atlantic continental shelf. The "live bottom" consists of hard and soft corals (near the northern limit of their geographical range), and is inhabited by a wide variety of fish and shellfish.

- Gulf of the Farallones NMS (desig. 1985), located north and west of San Francisco, encompasses the marine environment surrounding the Farallone Islands. Protected is a highly productive 948-square-mile area which is a habitat for several species of marine mammals and is the largest seabird rookery in the contiguous United States.

- *Monitor* NMS (desig. 1975), located off the coast of North Carolina, protects the sunken Civil War vessel USS *Monitor*, which has become an important marine habitat.

There are more NMS mandatory designations expected in the near future, and they include Monterey Bay, which will encompass a large area along the central California coast, and Western Washington Outer Coast, covering a large area along the Washington State coast. Additional proposed sanctuaries include Stellwagen Bank (off Provincetown, Massachusetts), Northern Puget Sound (in Washington State), and Norfolk Canyon (off the coasts of Virginia and North Carolina).

Numerous problems have plagued the marine sanctuaries program from the start, and it has never enjoyed the widespread public support that exists for terrestrially oriented parks and preserve programs. The most significant problem, from which all others follow, is that the program has never received adequate funding and support from either Congress or the executive branch. In fact, Congress did not appropriate any money for the program in the first seven years of its existence. Given the lack of funding, the program has not been able to fulfill its mandate. From a list of over 100 potential sites nominated for sanctuary designation in the late 1970s, NOAA has designated only those described above. Some sanctuaries have no resident manager, and research and education efforts have been widely curtailed. Substantial funding increases are needed in order to make this a truly effective program.

National Estuarine Research Reserve System

The mission of the National Estuarine Research Reserve System, administered by NOAA (under Section 315 of the Coastal Zone Management Act), is the establishment and management, through federal-state cooperation, of a national system of estuarine research reserves representative of the various biogeographical regions and estuarine types in the United States. While the major purpose of the research reserve system is the promotion and coordination of estuarine research, commercial development is prohibited or strictly controlled, thereby preserving biodiversity in discrete areas of the estuarine environment. NOAA notes that the primary intent of the system is the protection of natural pristine estuarine sites to provide opportunities for long-term research, education, and interpretation. However, restoration and recovery of degraded estuaries is also a function of the research reserve system.

A National Estuarine Research Reserve is defined as:

> an area that is a representative estuarine ecosystem suitable for long-term research, which may include all of the key land and water portion on an estuary, and adjacent transitional areas and uplands constituting to the extent feasible a natural unit, and which is set aside as a natural field laboratory to provide long-term opportunities for research, education, and interpretation on the ecological relationships within the area.[3]

The reserves are composed of a "core area" and a "buffer zone." The core area is the key land and water areas within the reserve; the area is so vital to the functioning of the estuarine ecosystem that it must be under a level of control sufficient to ensure the long-term viability of the reserve for research on natural resources. A buffer zone is an area adjacent to or surrounding the core area and essential to its integrity. The buffer zones serve to protect the core areas and to provide additional protection for estuarine-dependent species.

NOAA had designated research reserves in 17 areas by 1990, with several additional sites moving through the predesignation process. Estuarine reserves exist along all four coasts of the United States and also in Hawaii, and the 17 sites collectively total nearly 300,000 acres of estuarine waters, marshes, shorelines, and adjacent uplands.

National Wildlife Refuge System

The National Wildlife Refuge System consists of lands administered by the Department of the Interior's Fish and Wildlife Service specifically for the benefit of wildlife preservation. According to federal law, the system was created to conserve fish and wildlife, including species that are threatened with extinction. While predominantly a terrestrially oriented system, numerous wildlife refuges are located in coastal areas including coastal wetland, estuarine, and nearshore environments. Management goals of the refuge system are designed, among other things, to: (a) preserve, restore, and enhance in their natural ecosystems (when practicable) all species of animals and plants that are endangered or threatened with becoming endangered; (b) perpetuate migratory bird resources; (c) preserve biological diversity; and (d) provide the public with an understanding and appreciation of fish and wildlife ecology and the human role in the environment.

The Secretary of the Interior is allowed to permit any use within a refuge if it is determined that such use is compatible with the major purpose of the refuge. Unfortunately, this broad discretionary authority has allowed several commercial activities to be introduced into refuges. Currently, for example, the Arctic National

Wildlife Refuge in the Alaskan coastal plain is threatened with oil and gas development, and multiple-use restrictions have been the major issue in the ongoing debate surrounding a "wilderness" designation of that wildlife refuge. The *Exxon Valdez* oil spill in Prince William Sound raised the nation's consciousness about the vulnerability of pristine Alaskan coastal ecosystems to oil development activities. It is feared that the development of oil resources in the Arctic National Wildlife Refuge will severely endanger coastal ecosystems there.

Prevention of Overexploitation of Species

Endangered Species Act

The Endangered Species Act (ESA) is the most comprehensive federal law for the protection of species diversity and species habitats. The law gives to the Secretary of the Interior, acting through the U.S. Fish and Wildlife Service (FWS), responsibility for the recovery of terrestrial species and some marine species, and to the Secretary of Commerce, through the National Marine Fisheries Service (NMFS), responsibility for the recovery of most marine species.[4] The ESA authorizes the Secretaries to identify endangered or threatened species, designate habitats critical to their survival, establish and conduct programs for their recovery, enter into agreements with states, and assist other countries to conserve endangered and threatened species. The federal government is also authorized to enforce prohibitions against or issue permits controlling the taking of and trading in endangered and threatened species. Perhaps the most important provision of the law is that federal agencies are prohibited from funding, authorizing, or carrying out any projects that jeopardize the existence of or modify the habitats of endangered species.

An additional measure of protection to endangered and threatened species is the designation of "critical habitat." Under the act, critical habitat is defined as:

> (i) the specific areas within the geographical area occupied by the species . . . on which are found those physical or biologi-

cal features (I) essential to the conservation of the species and (II) which may require special management considerations or protection; and (ii) specific areas outside the geographical area occupied by the species upon a determination by the Secretary [of Interior or Commerce] that such areas are essential for the conservation of the species.[5]

While there are no inherent restrictions on human activities in an area designated as critical habitat, all activities that occur in that area are affected if they are authorized, funded, or carried out by federal agencies. Critical habitat designation notifies federal agencies that a listed species depends on a particular area for its survival and that any federally authorized activity may be prohibited if it is determined that such activity will adversely affect the species' habitat.

Implementation of the Endangered Species program has been controversial, and has been severely hampered by the lack of funds, especially during the last several years. Listing of marine species has lagged far behind that of terrestrial species: about 40 of the 340 species listed are marine, the majority being seabirds.[6] Only four of those marine species have critical habitat designations under the act.[7] The relative lack of marine species on the endangered and threatened species list is often interpreted to mean that marine species are not generally threatened. A more accurate interpretation would be that it is difficult to determine when marine species are threatened because their distributions are so poorly known.

Marine Mammal Protection Act

The Marine Mammal Protection Act (MMPA), passed in 1972, recognized the need to protect all marine mammal species, many of which were in danger of extinction or depletion as a result of human activities. The act was established to halt the decline of these species and to restore the populations to healthy levels. The National Marine Fisheries Service manages all cetaceans (whales and dolphins) and all pinnipeds except walruses (i.e., seals and sea lions). The U.S. Fish and Wildlife Service manages polar bears, walruses, sea otters, manatees, and dugongs.

The MMPA's primary management tool is, with several exceptions, a prohibition on the taking of marine mammals. "Taking" includes hunting, capturing, killing, and harassing. The act also prohibits most importations of marine mammals or their products. Several major exceptions, however, allow the killing or capturing of marine mammals. NMFS and FWS are authorized to grant permits for the taking of marine mammals for scientific research, public display in aquariums, subsistence purposes, and incidental takings during commercial fishing operations. Permits are allowed for these exceptions only after the appropriate federal agency examines the available scientific evidence and determines that the taking will not jeopardize the species.

The goal of the MMPA is to maintain marine mammal population levels at or above the "optimum sustainable population (OSP)." OSP is defined as the range of population levels from the largest supportable within the ecosystem (carrying capacity) to the population level that results in maximum net productivity.[8] If the population levels fall below OSP, the population is declared "depleted." With such a designation, intentional takings are permitted only for research purposes or for subsistence and handicraft purposes by Alaskan natives, and a species recovery plan must be developed. Species designated as endangered or threatened under the ESA are automatically designated "depleted" under the MMPA.[9]

Implementation of the act has been controversial, particularly with respect to the taking of marine mammals during commercial fishing operations. This has been especially true for tuna fishing and its effects on dolphin populations in the Eastern Tropical Pacific Ocean. Over 100,000 dolphins per year are killed during tuna-fishing operations. Other controversial issues include (a) the educational value versus entertainment value of marine mammals taken for aquariums, (b) preservation of marine mammal habitat, (c) entanglement of marine mammals in dormant fishing nets and plastics, and (d) direct harvesting of marine mammals for subsistence.

Fisheries Conservation and Management Act

With the enactment of the Magnuson Fisheries Conservation and Management Act (FCMA) in 1976, the United States extended its jurisdiction and control over all marine fisheries resources within 200 miles of the U.S. coast.[10] This action was taken largely in response to the increasing presence of foreign fleets, which were severely depleting many of the offshore species of fish. It is estimated that the U.S. 200-mile exclusive economic zone contains 20 percent of the world's fisheries resources.[11]

The FCMA established eight regional fishery management councils composed of state and federal fishery officials and industry representatives. The councils are charged with preparing, monitoring, and revising fishery management plans for each fishery of the EEZ that requires conservation and management. The plans must include measures to rebuild and restore fish stocks, prevent overfishing, and assure an optimum yield from each fishery. Common fishery management measures include seasonal restrictions, gear restrictions (e.g., size of mesh), closed areas, size limitations, and limiting entry to the fishery. Fishery management plans are approved by the Secretary of Commerce acting through NMFS. Coastal states maintain management control over fishery resources within state waters (three miles offshore for all states except Florida and Texas, where it is nine miles).

Foreign fishing, which dominated in waters offshore the United States in the late sixties and early seventies, has been virtually eliminated as a result of implementation of the FCMA. U.S. vessels, however, have more than compensated for the lack of foreign fishing. The ten most popular commercial species remain significantly below historical levels, with catches declining. In addition to overfishing, fisheries habitat destruction is playing a major role in the decline of some fisheries. Estuaries and coastal marshes, in particular, are under severe development pressure. This aspect of the problem, however, is largely out of the hands of fisheries managers in that agencies regulating coastal development are not the same agencies regulating fisheries management.

Water Quality and Habitat Protection

Clean Water Act

The Clean Water Act (CWA) is the primary federal legislation governing water pollution control. The primary objective of the act is to maintain and restore the chemical, physical, and biological integrity of U.S. waters. To accomplish this, Congress established a combined federal and state system of controls to implement clean water programs. The CWA consists of two major parts: (a) the federal grant program to help municipalities build sewage treatment plants, and (b) the pollution control programs, which consist of regulatory requirements that apply to industrial and municipal dischargers. The Environmental Protection Agency (EPA) has primary authority for implementing the provisions of the act, with substantial responsibility left to the states. For example, the EPA is responsible for developing water quality criteria, but it is up to the states individually to promulgate water quality standards based on the EPA criteria. Until standards are in place, they do not carry the power of the law. The act sets up a permit system for all discharge of pollutants into navigable waterways (the National Pollutant Discharge Elimination System, or NPDES) and gives the EPA and/or the states the jurisdiction over the system.

Three portions of the act are especially relevant to coastal and marine habitats and marine water quality: a provision for criteria applied to discharges into ocean waters, the regulation of disposal of dredge and fill material, and a program for protection of estuaries.

The act provides protection from direct discharges into marine waters through the application of the Ocean Discharge Criteria of section 403(c). Prior to issuing any NPDES permit for discharge into marine waters, the EPA must determine that the discharge will not "unreasonably degrade the marine environment." "Unreasonably degrade" in this case means: significant adverse changes in ecosystem diversity, productivity, and stability of the biological community within and surrounding the discharge area; threat to human health through direct exposure to pollutants or through consumption of exposed aquatic organisms; or loss of aesthetic, recreational, scientific, or economic values that is unreasonable in relation to the benefit derived from the discharge. These criteria, however, do not apply to waters shoreward of the territorial sea baseline and therefore do not apply to discharges into estuaries and other coastal waters such as the Chesapeake Bay and Puget Sound. In addition, the EPA only began receiving funding for this section in 1987.[12]

Section 404 of the CWA is the federal government's primary tool for controlling the use of wetlands and other coastal habitats through regulation of the discharge of dredge and fill material into the waters of the United States. The U.S. Army Corps of Engineers is the principal regulatory authority under the 404 program. The EPA, NMFS, and FWS have advisory roles in determining whether or not permits should be issued, although the EPA has the authority to overrule a decision of the Corps if it is determined that unacceptable impacts will occur as a result of the permit issuance.

The National Estuary Program is an important component of the CWA. The NEP was developed to address estuarine and coastal water pollution comprehensively by developing management plans for a number of specific estuaries. Ideally, the plans will address the control of point and nonpoint sources of pollution, implementation of environmentally sound land-use practices, the control of freshwater input and removal, and the protection of marine living resources.

Despite strides made in abating pollution in many rivers and lakes, coastal waters have steadily continued to decline. Numerous government reports document the degradation of marine and coastal habitat quality as a result of pollution. This degradation of coastal waters is due in part to the fact that the EPA has traditionally focused on the freshwater environment and devoted much less attention to the problems affecting coastal water quality. Few water quality standards have been enacted for coastal waters, a situation that may be rectified only through amendments to the CWA; amendments that would require the EPA to set national water quality standards instead of waiting for the states to do it. Another severe shortcoming of the implementation of the act is that biological diversity is not considered when a discharge permit is issued. Guidelines have not yet been developed that effectively address the impacts of pollution on reproduction, survival, and health of marine species and marine ecosystems.

Ocean Dumping Act

The Ocean Dumping Act (Title I of the Marine Protection Research and Sanctuaries Act) governs the disposal[13] of all materials into the ocean, including sewage sludge, industrial waste, and dredged materials—the dredge spoils by far accounting for the largest amount of material dumped into the ocean. The EPA and the Army Corps of Engineers are the permitting agencies for ocean dumping, while NOAA is responsible for research and monitoring and the U.S. Coast Guard for surveillance and enforcement. Disposal activities may occur only at sites officially designated by the EPA, and permits are issued for ocean dumping of certain materials if it is determined that such dumping "will not unreasonably degrade or endanger human health, welfare, or amenities, or the marine environment, ecological systems, or economic potentialities."[14]

The EPA must review each permit application based on specific factors including the need for the proposed dumping; its effect on human health, the environment and economic and recreational values; and, perhaps most importantly, an evaluation of alternative disposal options and their potential impacts. Certain materials are prohibited from being dumped into the ocean, such as high-level radioactive wastes, chemical and biological warfare agents, and medical wastes. In addition, recent amendments to the Ocean Dumping Act call for an end to ocean dumping of sewage sludge and industrial wastes by 1992.

While the regulation of ocean dumping may indirectly protect biological diversity, little or no attempt is being made by the federal government to assess direct impacts on diversity. Over 60 years of dumping sewage sludge and industrial waste off the New York–New Jersey coast along with pollution from land have caused extensive damage to the marine environment in the vicinity of the dump site and beyond, including a reduction in the diversity of benthic organisms now inhabiting the area.

Coastal Zone Management Act

The Coastal Zone Management Act provides federal grants to states to develop coastal zone management plans that balance the pressure for economic development with the need for environmental protection. Management policies for the coastal zone are ideally intended to protect coastal natural resources (including estuaries,

bays, beaches, and fish and wildlife and their habitat) and to encourage area management plans for estuaries, bays, and harbors.

Perhaps more than any other program, a coastal state's coastal
zone management program can contribute significantly to the protection of coastal ecosystems. Coastal states have the authority to
control land-use development in the coastal zone as well as most
water uses within a state's costal waters. The law also provides
states with authority to influence federally authorized activities by
ensuring that those activities are "consistent" with the state's
coastal zone management plan. The Coastal Zone Management
Program, however, has been beset with problems—ranging from
the lack of funding to battles over the scope of states' rights to influence federal activities.

Coastal Barrier Resources Act

The Coastal Barrier Resources Act seeks to protect coastal barriers
in the Atlantic Ocean and Gulf of Mexico by discouraging development. The act prohibits expenditures of federal money for development of infrastructure such as roads, sewer systems, water supply
systems, bridges, and jetties. Coastal barriers serve as the coast's
first line of defense against the ocean's tidal and wave energy, and
they typically protect highly diverse lagoons and wetlands. The act
established the Coastal Barrier Resources System (CBRS), which is
composed of thousands of miles of barrier islands along the Atlantic and Gulf coasts. The CBRS should be expanded to include the
West Coast and the Great Lakes shoreline. The act recognizes the
importance of protecting barriers and their shoreward estuaries
and associated wetlands, which are rich spawning, nursery, nesting, and feeding areas for economically important species and for
many threatened and endangered species.

Fish and Wildlife Coordination Act

The Fish and Wildlife Coordination Act (FWCA) is the major legislation under which the Fish and Wildlife Service and the National
Marine Fisheries conduct fish and wildlife conservation activities.

The law requires interagency coordination and consultation to assure that fish and wildlife resources and their habitats are considered in determining the impacts of federally funded or authorized projects that control, modify, or develop the nation's waters. The NMFS and FWS analyze a wide variety of permits under FWCA, including dredge and fill permit applications, hydroelectric power plant proposals, and diversion of fresh water for various projects.

The primary problem relating to the protection of marine biological diversity with respect to this statute is that the comments from either NMFS or FWS on federally authorized projects are only advisory, and thus are not necessarily incorporated into a project when it is implemented. Other federal agencies, such as the Army Corps of Engineers, are the permitting agencies and thus have final authority to issue the permits under the conditions they specify. Further, NMFS, FWS, and the EPA do not use biological diversity as a parameter for their recommendations, except insofar as officially designated endangered species or commercially or recreationally important species are involved.

Marine Plastics Pollution Research and Control Act

The problem of plastics pollution in the marine environment and the resulting impacts on fisheries and wildlife has come to the forefront of public attention over the last couple of years. Entanglement in plastic debris and drifting fish nets is a significant cause of death to marine animals. Particular attention has been focused on marine mammals and sea turtles (including endangered species) that have been choked, suffocated, and drowned by such debris. Congress responded by enacting the Marine Plastics Pollution Research and Control Act of 1987, which bans the dumping of plastics, including synthetic fishing nets, within the U.S. EEZ and by U.S. vessels anywhere in the ocean. Under an international agreement (MARPOL, Annex V; see Chapter 7), the United States and most other nations have agreed to ban the dumping of plastics by their vessels. The act also requires several studies to be conducted by the EPA and NOAA to determine the extent of the impacts of plastics pollution on fisheries and wildlife and to explore methods to reduce such waste in the marine environment.

Oil Pollution Act of 1990

In 1989, the oil spill of the *Exxon Valdez* in a pristine coastal area of Alaska offended the U.S. citizenry with the senseless and extensive loss of marine life that it caused in an area otherwise little touched by the impact of development. In response to this and a rapid series of other oil spills causing severe ecosystem damage in locations scattered around the world, Congress passed comprehensive oil-spill liability legislation in 1990. The Oil Pollution Act defined stiff terms of liability for such spills, designed to provide for rapid responses of cleanup and damage assessment and rapid payment of damages to those incurring loss or injury from an oil spill. Ultimately, the act may be protective of marine biological diversity by providing a financial deterrent that will foster stronger measures on the part of oil companies to prevent future oil spills.

Specifically, the new law requires that new oil tankers be built with double hulls and that all existing single-hull vessels be phased out by the year 2014. The law also increases federal liability limits for vessels from $150 per gross ton to $1,200 per gross ton per spill and creates a $1 billion federal oil spill fund, which is financed by the oil industry through a five-cent-a-barrel fee that will pay for damages when parties cannot be found or are unable to pay, or liability limits have been reached. The law requires review of licensed seamen's driver's license records along with drug and alcohol testing, establishes ten regional oil spill response groups, and requires federally approved oil spill contingency plans for vessels and oil facilities.

Comprehensive Environmental Response, Compensation, and Liability Act

Better known as Superfund, the Comprehensive Environmental Response, Compensation, and Liability Act (CERCLA) contains emergency response and cleanup provisions for chemical spills and for releases from hazardous waste facilities. The program has been used to identify numerous existing hazardous waste sites that need to be cleaned up. Many of these sites are in the coastal zone, some under water. There are several potential effects of this act on marine biological diversity. Many contaminated marine sites may eventually be cleaned up, and as the criteria for site designation are required to include effects on living marine resources as well as on human health, protection and cleanup will be afforded to greater areas of polluted coastal waters, especially estuaries. However, the act provides for possible ocean disposal of the contaminated wastes, which will threaten diversity in chosen disposal sites. Thus, for marine ecosystems, Superfund may be a double-edged sword.

National Environmental Policy Act

The National Environmental Policy Act (NEPA) requires that an environmental impact statement (EIS) be prepared for all proposed legislation and all major federal actions that could significantly affect environmental quality. Under the act, the EPA is exempt from this requirement, but that agency has formally adopted a policy of preparing such statements. The EIS requirement has proven important in the protection of the environment, because it provides the public with a written assessment of potential environmental impacts that can be critiqued and sent back for improvement if necessary. When done properly, the environmental impact statement provides information for making a responsible decision with regard to the proposed activity. Public interest groups can use the findings of an EIS (or its inadequacy, if that is the case) to formally protest a decision permitting an activity. To date, biological diversity as such has not been a parameter routinely assessed in EISs; however, that may change if biological diversity legislation currently proposed is eventually approved by Congress. Even without a legislative mandate, agencies preparing EISs may begin including biological diversity assessments.

Outer Continental Shelf Lands Act

The Outer Continental Shelf Lands Act was intended to encourage exploration and development of hard minerals, oil, and gas on the outer continental shelf within environmentally acceptable guidelines. The jurisdiction over this program lies within the Department of the Interior, whose responsibility it is to weigh the value (economic, social, and environmental) of a marine resource against the environmental impact expected from exploiting the resource. The Department of the Interior is authorized to lease parcels of the continental shelf and issue permits for marine mineral exploration. Parcels may be removed from the lease program and set aside as ecological reserves by act of the President. Marine Sanctuary designation does not automatically eliminate mineral exploration within those boundaries, but the President, through NOAA, can choose to do so within the context of each individual Marine Sanctuary designation. Congress is allowed to place one-year moratoriums on specified lease parcels and has used that privilege extensively since the inception of the program.

The development of oil and gas and hard minerals on the sea floor has the potential to cause great environmental damage. To begin with, the benthic ecosystems are totally disrupted in the vicinity of the activity, which can involve a very large area. In addition, the vast quantities of sediment released into surrounding waters can smother benthic communities and cloud the water column, with the result that an entire community structure is altered. Inevitably, there is a loss of species diversity in surviving communities within the area of impact. Also, when oil resources are developed there is a great risk of oil spills, which can spread out and destroy large areas of rich bottom communities and damage food chains and reproductive cycles in water-column communities. All these concerns have led Congress to impede the progress of this program until there is a better understanding of the true magnitude of the environmental threats.

To help alleviate this stalemate, President Bush in 1990 announced several decisions delaying—and in one case permanently prohibiting—oil and gas development in several areas off the U.S. coast. The announcement came as a result of a National Academy of Sciences panel determination that information is inadequate for making leasing decisions in areas off the coasts of California and Florida. The President canceled lease offerings off California, off southern Florida, in the North Atlantic, and off Washington and Oregon pending the completion of additional oceanographic and socioeconomic studies. Leasing in these areas will not be allowed before the year 2000 and then only if development is consistent with five "guiding principles":

1. Sufficient scientific and technical information must be available.

2. The environmental sensitivity of certain areas will preclude development, and development will be considered in areas where science and experience and new recovery technologies show development may be safe.

3. Priority for development will be given to those areas with greatest potential.

4. Energy requirements for the nation's economy and overall costs and benefits of various sources of energy must be considered in deciding whether to develop offshore oil and gas resources.

5. External events, such as supply disruptions, might require a reevaluation of the OCS program.

In addition, the President has prohibited the development of oil and gas within the Monterey Bay National Marine Sanctuary.[15]

However, these steps were not taken without a price. That price, unfortunately, appears to be Alaska and its offshore areas. Development of oil reserves in the Arctic National Wildlife Refuge and the Chukchi Sea will be supported forcefully by the Bush administration. While certainly not "the" answer to protecting sensitive offshore environments from the effects of oil and gas development, it is hoped that these decisions will go a long way toward bringing about positive changes to the OCS oil and gas program.

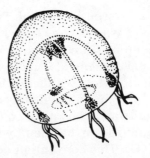

Chapter Seven

International Conventions and Programs

GIVEN THAT THE effects of human encroachment on the marine environment do not respect national borders, international action is necessary. However, there have been few international efforts to protect marine biological diversity. Nevertheless, there are numerous international treaties and programs that are designed to protect the environment. Several of those most relevant to the marine environment are discussed below. International conventions are binding only on the ratifying nations, but they must be honored by other nations when those nations operate within the territorial seas of any ratifying nation. When the United States ratifies an international convention, an act is customarily passed by Congress that directs the implementation of that convention.

London Dumping Convention

The Convention on the Prevention of Marine Pollution by Dumping of Wastes and Other Matter—commonly known as the London Dumping Convention, or the LDC—is the most significant international agreement dealing with the disposal of wastes at sea, including all waters within and beyond territorial seas. The LDC has been ratified by 65 countries, including the United States, and the administrative means of communication and cooperation among the signatories is fulfilled by the International Maritime Organization (IMO). This convention defines a "black list" of hazardous materials that are prohibited from ocean disposal and a "grey list" of materials that may be ocean-dumped only by special permit. (These lists are defined respectively in Annex I and Annex II of the convention.) The "black list" includes substances such as radioactive materials, biological and chemical warfare agents, persistent plastic oils, heavy metals like mercury and cadmium, and toxic organics like organohalogens. Substances that are not on either list require a general permit for dumping from the nation of origin of the vessel or of the material being dumped.

Although the threat of toxic pollution is greatest from land-based pipelines and runoff, the LDC is an important first step, significantly reducing the purposeful shipping of hazardous materials for disposal in the ocean. Effects on biological diversity have not been used directly as a criterion for establishing the "black" and "grey" lists, but the protection of diversity is implicit in such international regulations.

MARPOL

The International Convention for the Prevention of Pollution from Ships, better known as MARPOL, is aimed at reducing pollution from oceangoing vessels. There are five annexes, each dealing with separate regulations: Annex I, oil pollution from ships; Annex II, vessel discharge of noxious liquid substances in bulk; Annex III, release of harmful substances in packaged form; Annex IV, sewage; and Annex V, nondegradable plastics and garbage from ships. The

United States has ratified Annexes I, II, and V. The IMO, as with the London Dumping Convention, acts as the administrative branch of this international agreement.

The importance of this convention to marine life has previously been discussed. In addition to the concentrated disastrous effects of accidental oil spills, a great deal of harm is done by the small-scale discharges of oil from normal operations and the cleaning of vessels. In total volume of oil released into the sea, the small-scale discharges supersede the former by a great margin. Since MARPOL regulations were put in place, the incidences of oiled animals from such discharges have been noticeably reduced. A similar positive effect from the plastics regulations is expected over time, since the greatest source of that type of pollution is vessels.

Law of the Sea Convention

This is an ambitious treaty that has, to date, been ratified by 35 nations, not including the United States. A minimum of 60 signatories is required for the treaty to go into effect. Among the broad range of provisions that attempt to set up a system of international marine law are: the regulation of offshore resource development, pollution regulations and liability, scientific research, and marine mammal protection. If adopted, this convention could be the most significant law applying to the world oceans. Although biological diversity is not explicitly addressed by this convention, the far-reaching implications are obvious.

International Whaling Commission

The adoption of the International Convention for the Regulation of Whaling in 1946 established the International Whaling Commission (IWC). The IWC, whose membership now stands at 38 nations, regulates the hunting of whales by establishing catch and size limits and gear specifications, and disseminates information on whale biology. The original intent of the IWC was to regulate the harvest of whales for the benefit of the whaling industry.

The IWC failed disastrously in regulating the harvesting of whales, causing the stocks of all the great whales to plummet. With growing public sentiment toward whale conservation, the IWC changed from a "whalers' club" to, essentially, an intergovernmental whale conservation organization. Its vote for a moratorium on commercial whaling, which began in 1986, carried several notable abstentions—Japan, Norway, and the Soviet Union (the Soviet Union continued commercial whaling until 1988). Controversy over whaling continues, however, over so-called research whaling by Japan and Iceland. Whaling for research purposes is, in theory, allowed by the 1946 convention. However, the IWC has condemned the proposals for research whaling by these nations in the last three years as being a thinly veiled ruse for continued commercial whaling.

Despite international pressure, some nations have continued to practice commercial whaling. The IWC is powerless to enforce its resolutions and depends on international pressure and the enforcement policies of individual nations. Aboriginal subsistence whaling is still allowed by the IWC.

United Nations Environment Program

The United Nations Environment Program is the premier international organization dedicated to the protection of the environment in general and to the marine environment in particular. UNEP's Regional Seas Program seeks to develop unique regional arrangements to control pollution and manage marine resources. The Regional Seas Program encompasses 10 regional seas with over 120 nations participating.[1]

The Regional Seas Program is an action-oriented program encompassing a comprehensive approach to marine and coastal areas and to environmental problems concerning not only the consequences but also the causes of environmental degradation. Each regional program is shaped according to the needs of the nations in the region. The regional programs promote the parallel development of regional legal agreements and of program activities as embodied in action plans. UNEP serves a catalyzing and coordinating role in the development of regional conventions and action plans, while implementation is largely left to the individual nations.

The overall strategy followed by UNEP in the development of regional seas programs is: (a) the promotion of international and regional conventions, guidelines, and actions for the control of marine pollution and for the protection and management of aquatic resources; (b) the assessment of the state of marine pollution, of the sources and trends of this pollution, and of the impact of the pollution on human health, marine ecosystems, and amenities; (c) coordination of the efforts with regard to the environmental aspects of the protection, development, and management of marine and coastal resources; and (d) support for education and training efforts to make possible the full participation of developing countries in the protection, development, and management of marine and coastal resources.[2]

Two regional seas conventions provide good examples of how important protection of marine ecosystems can be achieved through regional international agreements:

- The South Pacific Regional Environmental Protection Convention regulates pollution from vessels, land-based sources, airborne sources, ocean dumping, storage of toxic or hazardous wastes, nuclear testing, exploration of seabed resources, and coastal erosion. It also sets up protective measures for coral reefs and wetlands, and addresses the problem of overfishing. It requires the ratification of at least 10 South Pacific nations before it can go into effect.

- The Convention for the Protection and Development of the Marine Environment of the Wider Caribbean Region—commonly called the Cartagena Convention—was drafted to ensure sound environmental management throughout the Caribbean area. The convention obligates parties to implement measures to control, reduce, and/or prevent marine pollution from ship discharge, ocean dumping of wastes and hazardous substances, land-based sources, seabed exploration, and atmospheric contamination. It also provides for specially protected areas, which would include coral reefs and wetlands.

Another UNEP program relevant to the protection of marine bio-

111

logical diversity is the joint UNEP/FAO Global Plan of Action for the Conservation, Management, and Utilization of Marine Mammals. A basic goal of the plan is to generate a consensus among governments of the world that would lead to a global policy for marine mammal conservation. Action Plan activities are organized into five categories: (a) policy formulation, (b) regulation and protective measures, (c) scientific research, (d) improvement of laws and their application, and (e) public awareness. The plan also incorporates 38 recommendations to deal with such subjects as the creation of sanctuaries, the prohibition of access to breeding grounds, the setting of catch limits, a review of fisheries interactions, and an evaluation of the effect of scientific sampling on protected stocks.

International Union for the Conservation for Nature and Natural Resources

The mission of the International Union for the Conservation for Nature and Natural Resources (IUCN) is "to provide independent international leadership for promoting effective conservation of nature and natural resources."[3] Its membership consists of national governments, government agencies, and nongovernmental organizations. IUCN conducts numerous projects related to biological diversity around the world, many of which are related to the marine environment.

The principal manifesto for the IUCN, as stated in its published report *World Conservation Strategy (WCS)*, is to help advance the achievement of sustainable development through conservation of living resources. The three main objectives of the manifesto are: (a) to maintain essential ecological processes and life-support systems, (b) to preserve genetic diversity, and (c) to ensure the sustainable utilization of species and ecosystems.[4] IUCN is in the process of developing a second edition of the *WCS* (*WCS II*), which will include a new chapter on marine conservation strategies.

While not a major program within the context of the overall budget of the IUCN, the Coastal and Marine Conservation Program calls for, among other things: (a) the preparation of a practical guide on operational units of the global coastal zone for conservation; (b) the preparation of a handbook for managing coastal units; (c) the production of an authoritative data set on the keystone species in each operation unit for monitoring purposes; (d) identification of gaps in the coverage of protected areas in the coastal zone; and (e) the development of a data base on marine species.[5]

Antarctic Treaty Programs

The Southern Ocean, the sea surrounding the Antarctic continent, is one of the most biologically productive marine regions. Krill is the so-called linchpin of the Antarctic food web in that it supports a wide variety of animal life including squid, fish, seals, whales, and seabirds. Despite the remoteness of the Antarctic region, massive hunting of whales, seals, and Antarctic cod have had untold effects on the ecosystem. Both sealing and whaling are now under control, yet potential new threats face the Southern Ocean.

In 1959, 12 nations adopted the Antarctic Treaty in order to deal with the potential use of the Antarctic continent for military purposes. This treaty defers all claims of sovereignty to lands and waters of Antarctica, prohibits use of the continent for military purposes and nuclear testing, and promotes the freedom of scientific research. Since then, natural resources have become the dominant international political force in Antarctica, and several new treaties have been adopted to address these issues.

The Convention for the Conservation of Antarctic Seals prohibits sealing on the open ocean and provides a regulatory framework for regulating other commercial sealing if it is initiated in the future. The Convention for the Conservation of Antarctic Marine Living Resources (CCAMLR) calls for an ecosystem approach to managing the living marine resources of the Antarctic ecosystem. CCAMLR is the first such international convention to be based on ecology and not on political boundaries. The area of responsibility of CCAMLR is the Antarctic convergence—a natural, physical-ecological boundary coinciding with the Antarctic circumpolar current. The convention was adopted because of the prospect of large-scale commercial harvesting of krill. CCAMLR stresses the need to adjust harvesting of krill so as not to impair the balance of the supporting components of the ecosystem.[6]

The most recent Antarctic treaty, the Convention on the Regulation of Antarctic Mineral Resource Activities, deals with the exploration and exploitation of Antarctic minerals. Oil and gas and hard minerals have not yet been discovered in Antarctica, and current market conditions prevent any industry interest in exploration for such minerals. Nevertheless, 33 nations adopted the convention to establish a system in the event minerals are discovered. It establishes a regulatory framework that specifies binding legal obligations with which all stages of minerals activities must comply.

Environmental groups and several nations have opposed this convention on the grounds that it regulates an activity that they believe should be prohibited. They would prefer to see a treaty that sets up Antarctica as an international protected wilderness where resource exploitation is prohibited.

U.S. Agency for International Development

The biological diversity program of the U.S. Agency for International Development (AID) represents the most extensive U.S. government effort to conserve biological diversity in the developing world. Under direction from Congress, AID has focused its conservation activities on:

- helping countries protect and maintain wildlife habitats and develop sound wildlife management and plant conservation programs;

- establishing and maintaining wildlife sanctuaries, reserves, and parks;

- identifying, studying, and cataloging animal and plant species; and

- assisting countries to enact and enforce antipoaching measures.[7]

While nearly $5 million was dedicated to the preservation of biological diversity in developing nations by AID during fiscal year

1987, only approximately $500,000 was spent for projects involving the marine environment.[8] Clearly there is a greater need for U.S. expenditures on such projects in developing nations.

Biosphere Reserve Program

Biosphere reserves are an established network of representative protected ecosystems around the world sponsored by the United Nations Educational, Scientific, and Cultural Organization. These protected areas, however, differ from the traditional concept in that the areas are based on the zoning of different uses, and attempts are made to incorporate the local people's needs into the planning and management process of the reserves.

In an idealized model, the heart of a biosphere reserve is a core area representing one of the earth's major ecosystems which is strictly protected and maintained free of human disruption.[9] Surrounding the core area are areas managed for agriculture, grazing, forest production, fisheries, recreation, or other economic uses of renewable resources. The emphasis in these areas is on sustainable development with minimal impacts on the environment.

In practice, however, biosphere reserves have not lived up to the ideal model. The majority of reserves are superimposed on existing protected areas, and adherence to the goals of the biosphere reserve program by managing agencies is voluntary.[10]

Terrestrial reserves by far dominate the system of biosphere reserves. While several coastal and marine biosphere reserves have been designated, the concept is not directly applicable to the marine environment.[11] Work has been initiated to tailor the criteria for biosphere reserve development to marine areas.

Chapter Eight ■■■■■■
Conserving Marine Biodiversity in the Future

THE PROBLEMS created by multiple uses of a fluid environment that cannot be compartmentalized and does not honor political boundaries are complex. The solutions must be innovative. We cannot simply transfer terrestrial environmental protection methodology into the ocean and expect it to work the same way. Since the major threats to marine biological diversity come from the land and the atmosphere, it will not do merely to draw imaginary lines around specific areas of marine habitat and expect everything within those boundaries to be protected. A more comprehensive approach is needed, one that involves regulation of human activities on land as well as in the ocean and regulation outside reserve borders as well as within. Toward this goal, both knowledge and innovation in many fields will be required. Science and technology must be used to develop nondestructive interactions between humans and natural ecosystems. Environmental regulations with

116

teeth will have to become politically more acceptable. A new economics that does not place such disproportionately high value on exploitation of natural resources will have to be developed.

It is also important that there be channels through which the public can receive and provide information with respect to the management of marine ecosystems. The importance of citizen pressure in driving the process should not be underestimated.

The following discussion addresses more specific recommendations for what is needed with respect to understanding, monitoring, and protecting biological diversity in the marine environment. As with all environmental problems, the support of governments, international institutions, and private citizens will be needed to implement effective solutions.

Research

Government support for scientific research on ocean ecosystems must be greatly increased to address the many unanswered questions relevant to marine biological diversity. This support is needed from governments of all nations. It can come in the form of research conducted within governmental institutions and agencies or grants awarded for research in academic institutions. It can also take the form of foreign aid to promote research on biological diversity in other countries through programs such as those sponsored by AID and MAB (UNESCO's Man and the Biosphere Program).

Oceanographic research may be able to provide the necessary information on the ocean's role in climate change, in the oxygen and carbon cycles, and in other geochemical cycles affecting the composition of the atmosphere. We need to learn more about how the ocean may influence and respond to global warming so that the models of climate change can be improved.

We must also learn more about the processes that regulate biological diversity in oceanic and coastal ecosystems and how pollution and harvesting impact those processes. Coastal ecological research is most effective on a regional or whole-ecosystem basis; therefore, regional research programs, often involving several countries, are needed to support the movement toward management of large marine ecosystems. The Regional Seas Program of UNEP (discussed in Chapter 7) offers a framework for cooperative research among countries that border designated coastal ecosystems. In the commons of the central ocean what is called for are international research programs with an ecological focus to study large ecosystems such as the Pacific gyres, the Gulf Stream, and the equatorial regions of the Atlantic and Pacific oceans. The governments of wealthy countries should provide the major funding for these cooperative research projects. Programs such as UNEP's International Geosphere and Biosphere Program, though not restricted to marine ecosystems, could offer significant opportunity for cooperative international research relevant to marine biological diversity—if it were adequately funded.

Besides oceanography, there is another important area of research that must be supported. A combination of systematics, taxonomy, and genetic research will provide a better understanding of the biological structure of marine communities. There are many unidentified species, especially in deep-ocean areas, and the distribution of many known species remains uncertain. Geneticists and population biologists are needed to sort out the genetic differences between populations of individual species, and the significance of these distinct populations to biological diversity. As is true for terrestrial biology, there is a dearth of systematists and taxonomists, because for a long time that area of research has not been popular among funding agencies. Systematics has taken on the aspect of a dying art, and it is critical that government funding begin to breathe new life into the field. Faculties of many university biology departments no longer include systematists, and marine biology programs frequently do not emphasize the importance of taxonomy as a critical tool for understanding the biological processes within marine ecosystems.

In addition to the disciplinary research described above, there is a great need for effective long-term monitoring of marine ecosystems. Until there are more long-term records of the structure of biological communities in the different types of marine ecosystems, it is impossible to assess human-induced impacts on those systems with any accuracy. An example of the kind of record that is desirable is the California Cooperative Oceanic Fisheries Investigations (CalCOFI) program, which has been monitoring the California Current ecosystem since 1949, representing one of the most complete ocean time series in the world. Very few marine ecosystems have been studied in as much detail over a long time. Monitoring is expensive and requires a great deal of careful planning if the results are to be meaningful, but the effort and expense are necessary in order to assess long-term impacts on marine ecosystems and their biological diversity.

Finally, there are important new technologies that in the future will greatly enhance our ability to understand ocean biology and to monitor changes in marine ecosystems. The first is remote sensing by satellite. The whole-globe image of earth as a single integrated system evolved over the past decade primarily as a result of images of earth from space and the new ability to monitor the earth from satellites. The other technology is manned submersibles, which allow for exploration of the deep-ocean floor.

Satellite data has revolutionized the analysis of ocean circulation by providing oceanographers with more thorough coverage of the ocean surface and more extensive time series analyses (analyses of quantitative changes in measurements over time) than ever before imaginable. Open-ocean biology has also been aided by satellites—in particular by color-scanning data, which has enabled assessments of phytoplankton distribution over space and time. Of course, satellites are good only for surface or near-surface observations. To study deep-water biology, other techniques are necessary.

The popularization of scuba diving in the fifties was an important breakthrough for marine scientists studying underwater communities near the coast. Now, with the more recent development of manned submersibles, deep-bottom communities (and anything on the way down) can be observed directly. Since the vessels can be equipped with remotely operated mechanical arms for collecting specimens they are invaluable for observation and description of underwater communities. The small one-person submersible holds great promise for deep-water research. It is easily launched because it is light, is highly maneuverable, and is relatively inexpensive to purchase and operate. Another important tool is the unmanned remotely operated vehicle, or ROV, which can be very small, since it only has to be large enough to house video and audio equipment and to accommodate mechanical arms. The ROV broadcasts television images to a ship on the surface.

All of these undersea exploration technologies share one major limitation: they can cover only very small areas. That problem may be solved in the near future, however. There is a project under way to build a one-person submersible that would sail through the deep ocean as a glider sails through the air.[1]

National and International Policy on Marine Biodiversity

Spurred on by the rapid demolition of species in the tropical rain forests, many nations have begun to reexamine national and international policies on biological diversity. It is now widely recognized that the endangered species approach is only partially effective in protecting diversity. In the face of rapid habitat destruction, the process of identifying endangered species cannot keep pace with the losses. Endless numbers of species are thought to have gone extinct in the felled rain forests before they could be identified. Marine ecosystems are even more vulnerable to the same inadequate assessments of species.

The solution now widely accepted for better protection of biological diversity is habitat protection, in the form of national and international parks, wildlife refuges, and other protected areas, including marine reserves. The designation of marine protected areas both by nations and by international agreement is still in its infancy; there is the problem of determining what maritime activities can be and should be restricted. The larger and deeper the area, the greater the problem, because issues like fishing rights, energy development rights, and international vessel traffic protocols hinder true protection of the biological reserve. Perhaps one of the more effective types of protected areas would be fish reserves, where important breeding grounds for commercial fisheries species are held off limits to fishing and other potentially harmful activities. Coral reefs may also benefit from the establishment of strictly regulated marine reserves, since they typically harbor an abundance of species that stay close to home. For most marine environments—including coral reefs subjected to pollution from land—the designation of reserves is not enough, and this approach should not be seen as the major means of protecting biological diversity in the oceans.

Consequently, any comprehensive national and international policy on marine biological diversity will have to include integrated coastal zone management and regulations affecting basic processes that ultimately impact coastal and oceanic waters—regulations affecting land use, water quality, and air quality, for example. Currently, national and international policy on biological diversity is being drafted with the terrestrial model in mind. Although meant to include all environments, the emphasis is invariably on the protection of endangered species and critical habitats, which will tend to benefit terrestrial species. An international convention on biological diversity is being drafted, and early drafts emphasize threatened and endangered species and protected areas. There are provisions beyond that which would benefit ocean systems, but, because of the emphasis, the convention may be less protective of diversity in marine ecosystems than in terrestrial ecosystems.

As mentioned previously, the focus on species as the critical unit of diversity may not be appropriate for marine ecosystems; measures of functional diversity, for example, may provide a more meaningful assessment of biological diversity. It may be that benthic marine ecosystems should be evaluated differently from pelagic systems. Whatever the important parameters and complexities involved in assessing and protecting marine biological diversity, it is clear that current national and international policy has not taken them into account. It is equally clear that policy makers need advice from the scientific community in this matter and that the scientists' recommendations will evolve as research provides more insights into the dynamics of marine ecosystems.

Marine scientists, policy analysts, economists, and environmentalists are beginning to address the problems of biological diversity in marine ecosystems (for example, a working group on marine biological diversity was convened at the Marine Policy Center of the Woods Hole Oceanographic Institution in 1989, and a special session on marine biological diversity was held at the 1990 annual meeting of the American Association for the Advancement of Science).

What is needed is an international focus on the threats that coastal urbanization and development present to diversity in marine ecosystems and a comprehensive world strategy on the conservation of marine biological diversity. A step in that direction is being taken by UNEP, IUCN, and the World Resources Institute, which, in cooperation with a number of national and international governments and organizations, have a program under way to develop a global conservation strategy to keep losses of biological diversity to a minimum and to manage living resources wisely.

International policies with respect to marine ecosystems and their biological diversity are of particular importance in marine environments where ocean currents sweep pollution and debris across political borders and where the open ocean is a commons that can be regulated only by international treaties. While certain estuaries and coastal waters may benefit from the enforcement of a protective national policy, regional seas bordered by numerous nations will be protected only through conservation agreements among all the nations. Strong international treaties and conventions will be critical in the future of marine conservation. Of course, international policy is effective only when enough countries agree to participate and when the participating nations aggressively implement the policy.

Sustainable Development

A proposed international policy that is also being currently debated is that of *sustainable development*. As proposed by the World Commission on the Environment and Development,[2] such a policy would apply to all future development, including that affecting marine environments. The concept is based on the realization that our planet's natural resources are finite and that it would be prudent policy to regulate our activities so that natural resources are maintained and not depleted. For self-replenishing (i.e., living) resources such as trees and fish, this means regulating the amount and location of the harvest and protecting habitat as to guarantee

122

that populations will be maintained. For nonreplenishing resources such as minerals and oil, the problem is more complicated. Recycling policies can alleviate the wasteful loss of some
minerals, but what about resources that are totally consumed, like
oil? Here the prudent policy would be energy conservation and the
development of alternate energy sources. Although this policy
would be categorized as energy policy, the effect on marine ecosystems is obvious: drilling and oil transport would be decreased, reducing the threat to marine biological diversity.

Also falling within the concept of sustainable development is sustainable agriculture. A policy applied to crop lands would have a
potentially remarkable impact on marine ecosystems that are now
heavily polluted by agricultural runoff. Countless estuaries and
semiclosed seas (e.g., the Gulf of Mexico) would benefit tremen-
dously from such a land-based policy.

The goal of sustainable use of natural resources seems inherently
sensible—possibly even attainable through heroic efforts and
reordering of priorities. There are, however, two major obstacles
to achieving sustainable development—human population growth,
which in and of itself will place ever-increasing demands on natural resources, and the current perception of what "development"
means. Development in the "developed" world has been closely
tied to technological advances, and the wealth and consumerism
that have resulted are the envy of "underdeveloped," or "developing," countries. As long as the aim of development is to achieve
ever-increasing wealth, promoting escalating levels of consumption of resources and products, sustainable development is a self-
contradictory goal and can never be achieved.

A sobering example of technological development for profiteer-
ing gone awry is the development of drift nets as a technique for
increasing the catch of North and South Pacific fisheries. Huge
nets often several miles long are let loose to drift in the ocean currents for indefinite periods of time, trapping anything too large to
pass through the mesh. In a scenario resembling science fiction,
these deadly oceanic gossamers float slowly through the sea collecting their varied harvests until each becomes a veritable sam-
pler of pelagic species diversity. To add insult to injury, many of
these nets are never recovered. That some people's standard of
living is enhanced by such wanton devastation of marine commu-

nities is humiliating. Through recent international pressure, the drift-net fleets of Taiwan, Japan, and South Korea have agreed to a ban on such practices in the South Pacific. No agreement, however, has been reached on limiting drift-net fishing in the North Pacific, where it is believed that fishing activity is much greater. The Atlantic Ocean is now experiencing this deadly technology.

On the other side of the marine environmental fence is technological development that is protective of marine ecosystems, such as the development of sewage treatment processes. This technology is used to protect human health rather than to reap large profits, however. Technological developments of this sort, enabling human society to live within the bounds of the natural world, should be encouraged.

Since it does not seem likely that the profit motive is going to drive development in the direction of sustainable use of natural resources, there will have to be either a redefinition of development goals or strict restrictions placed on certain types of development.

Economic Value of Marine Biological Diversity

Historically, economic value has been assigned to natural resources solely on the basis of marketability. Now economists are beginning to develop systems for assigning dollar value to natural resources that are not directly marketable, such as biological diversity, critical habitat, buffer zones, and hydrologically critical areas. A system of dollar values assigned to biological resources or biological diversity on the basis of both market and nonmarket values can be applied to living marine resources. One such classification system includes *direct values* (consumptive uses and productive uses) and *indirect values* (nonconsumptive uses, future options, and mere existence).[3] Direct values are assigned to specific resources, usually individual species or discrete groups of species, that are consumed; and indirect values are assigned primarily to ecosystems or ecosystem functions that have some perceived value to society. Such a classification system allows more realistic cost-

124

benefit analyses of activities affecting species and ecosystems. In some cases, the indirect value of an ecosystem will outweigh the direct value of harvesting some species or group of species when that harvesting threatens to destroy or devalue the ecosystem.

A *consumptive-use value* is assigned to a particular biological resource that is consumed directly without going through the marketplace. In coastal environments, this includes fish, shellfish, sea mammals, and seaweeds harvested and used directly as food (i.e., recreational and subsistence fisheries); mangroves cut for firewood; submerged grasses and marsh grasses used for weaving objects used by native tribes; and species used by natives for medicinal purposes. The resources used are usually categorized as species or groups of species (e.g., salmon, mussels, dulse, sea grass) and values assigned do not reflect any exchange of money, but in some cases they may be compared to the market value of the resource. In the case of recreational fisheries, the pleasure of the activity is assigned a value in addition to the food value of the fish caught. Some marine resources with no market value can now be appraised on the basis of direct consumptive value.

A *productive-use value* is assigned to a biological resource that is marketed. Productive-use values have usually been the only values assigned to natural resources (for example, when a large forest fire occurs in a national forest in this country, the economic loss is always given in terms of board feet of lumber). Marine examples are obvious: commercial fisheries and aquaculture food species, species used for production of medicines, marketable oils of sea animals, species used in marketed crafts, building materials (as from coral), live specimens marketable for aquariums, seashells sold for decorative uses. Since these resources are given monetary value on the open market, there is no innovation necessary to derive the appropriate values.

A *nonconsumptive-use value* is assigned to an environmental resource that provides a service (ecosystem function) that does not necessarily have a market value but provides obvious benefits to society. For marine ecosystems, these beneficial functions include the following:

- buffering between the land and the sea, which reduces negative impacts in both directions; for example, barrier beaches absorb

storm energy, protecting the main coastline, and coastal wetlands absorb and break down pollutants from land that would
otherwise wash into the sea

- nursery grounds and/or feeding grounds for species having
  direct-use value
- primary productivity that enters food chains and supports commercially valuable species; for example, primary production in
  salt marshes and mangroves supports offshore fisheries
- important habitat for rare or endangered species
- storage and cycling of essential nutrients such as nitrogen and
  phosphorus
- role in bio-geochemical cycles, such as oxygen, carbon dioxide,
  and sulfur cycles
- role in regulation of local and/or global climate
- recreational and aesthetic values for tourism
- usefulness in scientific field research and education
- cultural value—historical, ethnic, or religious.

The value of these functions is easier to assign on a local level (e.g.,
a specific wetland is a valuable nursery ground for a specific offshore fisheries species) than on a global scale (e.g., open-ocean
systems play a significant role in the carbon cycle).

Option value is assigned to particular species or a level of diversity of species on the basis of potential future uses—i.e., keeping
options open in the face of an uncertain future. For example, it is
reasonable to assume that some marine species, perhaps as yet unknown, will be valuable in the future as sources of medicine, genetic resources for biotechnology, or the like; that some coastal area,
if protected, will have future value as a tourist spot; or that conserving the large genetic pool associated with a high level of biological diversity will prove important in the adaptation of living
systems to future environmental changes. It is difficult to imagine
how dollar values can be assigned to such options, but economists
are grappling with the problem, since it is clear that potential value
may become real value.

Existence value is assigned to species or habitats or ecosystems
in the hope that they may be of some benefit to future generations
or that there is some benefit derived from the mere existence of

other species on earth. For example, some people feel that we have a moral responsibility not to destroy species or ecosystems—not to destroy sea otters and whales, Prince William Sound and Antarctica. This is more an ethical value than a utilitarian one, and it is nearly impossible to convert this value into dollar terms. Some indication of value may be derived from an assessment of voluntary contributions made to private conservation organizations. This, however, does not reflect the value to people who do not have excess personal funds to donate to such organizations. The magnitude of existence value is destined to be a major bone of contention in cost-benefit analyses that incorporate this proposed system of assigning direct and indirect values to biological resources.

Insofar as economics seems to be the most powerful force driving human societies to modify, rearrange, or destroy ecosystems in their quest for space and natural resources, this new way of assigning value to biological diversity is to be encouraged. The move away from a singular focus on market values to a multiple value system, which includes human health, quality of life, and ethical considerations, holds great promise for ecologically sound cost-benefit analyses of proposed development projects in the future. Nevertheless, it does not eliminate the pitfall we humans always construct—our incessant need to view the natural world in terms of its value to human society. That anthropocentric perspective may yet prove to be the Achilles' heel of global conservation efforts.

Measures to Protect Biological Diversity

Once national and international policies protective of marine environments and their associated biological diversity are in place, the difficult and sometimes expensive job of implementing those policies begins. The long list of what is needed for effective implementation can be broken down into five categories: regulation of land-based and maritime sources of pollution; coastal zone management; direct regulation of marine resources; establishment of marine protected areas; and use of economic incentives and disincentives.

Regulation of Land-Based Sources of Pollution

Regulating land-based and maritime sources of toxic pollution may be one of the most effective ways of protecting marine biological diversity. Although it does not provide for direct protection of species, it attacks the single greatest ubiquitous threat to marine species and ecosystems. The key elements of such regulation are standards for water quality, sediment quality, and air quality, and standards for discharge of pollutants. Just how effective the standards are will be determined by scientists and regulators, but the standards must have some biological validity. Therefore, their effectiveness in protecting the marine environment should be measured by looking at biological components of affected marine ecosystems. Specifically, biological diversity should be used as an indicator.

The at-sea discharge of pollutants has been better regulated than land-based pollution at both the national and international levels. However, restrictions on the intentional discharge of hazardous wastes range from very strict—the oceanic disposal of high level radioactive wastes is prohibited, for example—to very weak. The disposal of dredge spoils, which include sediments contaminated with hazardous materials, is one area that has been neglected and needs attention because of the magnitude of the disposal problem. Although effective regulation of dredging and dredged material disposal is necessary, the long-term solution to contaminated sediments gets back to eliminating land-based sources of toxic materials.

Besides intentional dumping of wastes in the ocean, there is the serious problem of accidental spills of hazardous materials such as oil. Strong national and international laws regulating the transport of these materials and assigning liability could act as a deterrent to spills. Avoiding spills is far more important than providing for their cleanup, which is virtually ineffective in protecting marine ecosystems from severe long-term damage.

Where regulations have been established to reduce the threat to a particular marine ecosystem (a regional sea, for example), those regulations should be tested by the monitoring of biological diversity. Currently, concentrations of contaminants in the water are typically monitored for this purpose. Biological effects are less commonly used as indicators, and very rare indeed is the use of biological community structure or ecosystem processes as indicators.

Whereas the motivation to regulate land-based sources of pollution in the marine environment may come from international agreements, the regulations themselves will be established by each nation. Treaties may provide a framework for integration among the various national programs, but aggressive implementation at the national level is essential to making the treaties work. Mechanisms for enforcing international environmental treaties and conventions have historically been weak or absent. That needs to change if we are ever to achieve effective global coordination of regulations to control pollution in the world's oceans.

Coastal Zone Management

In light of the rapidly increasing pressures on the coastal zone as a result of the concentration of exploding human populations and development along or near the coasts, the importance of coastal zone management in protecting marine biological diversity is clear. Population increases give rise to urbanization; to increasing demands for energy, water, and building materials; to increased need for waste disposal facilities; and to poverty. At the same time, economic development and industrialization give rise to wealth and poverty, consumerism, tourism, and pollution.

Environmental protection of the coastal zone in areas under extreme population and development pressures often comes down to a question of intense management, since it is often unrealistic to set aside totally protected areas. Integrated resource management for the coastal zone is becoming widely accepted as the most effective means of dealing with the multitude of coastal problems related to development.[4] This approach recognizes the interrelated functions of the coastal zone and the need to get away from segregated "sector" management, where particular activities are regulated without taking into account other activities affecting the same resources.

The sector approach has been common, not only among managers but also among development lending agencies such as the World Bank. Problems arise when development projects are approved without a consideration of all the impacts on environmental processes and on other human activities taking place within the entire affected area. One activity may give rise to cascading effects as direct impacts on environmental processes are indirectly transferred to other interrelated processes. The impact is passed from one ecosystem to another throughout the coastal zone, mostly because of the basic properties of water (it flows, it carries a particulate load, and it dissolves).

For example, building a dam may have direct economic advantages, but the impact on activities such as farming and fishing upstream and downstream of the dam may be devastating. Upstream farmland is flooded and downstream farmland is deprived of nutrient-rich silt replenishment. The distribution of fish species is changed, and migration and breeding patterns are interrupted. Furthermore, estuarine resources far downstream may be dramatically affected by the change in freshwater inflow, and coastal surface and groundwater supplies may be damaged by saltwater intrusion. There may also be increased impacts from water pollution as the pattern and volume of water flow change. Biological diversity in several ecosystems may be affected—the river itself; the floodplain; associated wetlands; the estuary, where the river flows into the sea; the offshore and terrestrial ecosystems that depend on the river, its estuary, and its wetlands as a source of food and as a nursery area; and, of course, those dry-land areas that are permanently flooded by the new reservoir of water held by the dam.

Integrated resource management on a regional scale offers a great deal of promise, and those interested in such management on an international scale are beginning to define what is needed in the design and implementation of integrated coastal resources management programs.[5] It is now imperative that local regulators and managers, development banks and borrowing governments, understand and accept the wisdom of this approach, since strong institutional frameworks on national and regional levels will be necessary in order to turn theory into action.

Economic Incentives and Disincentives

Often it is not enough to establish a policy protective of biological diversity followed by research, laws, and regulatory and management institutions that are supportive of the policy. When these measures fail to maintain the level of biological diversity necessary for the welfare of the planet and human society, economic methods may provide an effective alternative. Economic incentives (rewards) or disincentives (punishments) may be implemented to encourage behavior protective of biological resources or to discourage destructive behavior. McNeely offers a definitive and detailed account of such methods at the community, national, and international levels, and anyone interested in the application of economics to the conservation of biological diversity should refer to that work.[6]

Up until now, governments have used economic incentives to encourage the exploitation of natural resources, including biological resources, often well beyond their capacity to replenish themselves. Now it is imperative that those economic policies be altered and that the sustainable use of biological resources become the guiding principle for economic incentives. Developing nations have an opportunity to redirect their development goals toward the maintenance of a critical level of biological diversity, and it is imperative that the developed nations encourage and abide by such goals through their international economic aid programs. The developed nations stand to gain great benefit from successful conservation efforts in developing countries.

At the same time, they should be instituting similar economic incentives at home. Industrialized parts of the world have already lost the opportunity to save much of their original biological wealth, but there is no need for them to continue exploitation at a rate that ensures the demise of those remaining natural ecosystems. Both to set an example and to acknowledge the importance of the natural world to their own societies, these countries need to revamp their economic guidelines. A system of economic rewards and punishments should be established by governments in the United States, Canada, Europe, the Soviet Union, Australia, and Ja-

131

pan to reduce pollution, deforestation, overfishing, and critical habitat destruction.

Specific incentives that could be implemented in the coastal zone include:

- subsidies for not building on coastal land in general (a one-mile buffer, for example) and excise taxes to build on barrier islands and other vulnerable coastline sites that are not already protected by zoning restrictions
- stiff fines for atmospheric, ground-level, or underwater discharges of pollutants exceeding specified levels (designed to make it economically prudent for industries to modernize using antipollution equipment) and tax breaks for discharges achieving even more stringent standards, with the idea that incentives will be more effective than current regulations and permitting procedures
- subsidies to fishermen for not harvesting certain fish that are being depleted (similar to subsidies for farmers not to grow certain crops or cultivate certain acreage).

While cleaning up their act at home, the wealthy nations of the world should also apply high standards of natural resource conservation to their international lending and financial aid practices. It is important that the lending countries stop trying to mold developing countries to their own model and that they start encouraging those countries to develop their own models, models that include integrated resource management appropriate for their own cultures and geography. For starters, the governing boards of the multilateral lending institutions should establish new guidelines for lending to turn previously negative impacts of their lending policies into positive incentives promoting sustainable use of living resources in the borrowing countries. Economic incentives through international lending should include the direct funding of conservation projects as well as the application of strict environmental standards to development projects being funded.

The U.S. Treasury Department has begun this task by issuing environmental voting instructions for proposed projects affecting tropical rain forests, wetlands, and coastal marine ecosystems. These instructions are based on background papers and recommendations prepared by committees of experts and environmental

132

organizations. The coastal marine ecosystems guidelines were developed on the basis of many of the principles that have been outlined in this book. Once issued by the U.S. Treasury Department, these guidelines are used to instruct and evaluate U.S. members of the executive boards of the multilateral development banks when they vote on lending money for a proposed development project in a developing country. The United States has also proposed that these voting instructions be adopted by the bank governing boards—so far without success.

Regulation of Marine Resources

The regulation of fisheries and other harvesting activities in the ocean is becoming more and more critical as overfishing by commercial and sport interests becomes more prevalent in these environments. Coastal waters, to a distance of 200 miles offshore, may be regulated by individual nations, and some regulation of this kind is in effect already, though not on the scale needed.

Coral reef areas are direly threatened by overharvesting and subsequent loss of species diversity as a result of destructive fishing methods and harvesting for the souvenir and aquarium trade. Stringent regulations must be established by the governments having jurisdiction over these threatened reefs. Trade restrictions can help, as in the case of the Convention on International Trade in Endangered Species of Wild Fauna and Flora (CITES). This treaty regulates trade of many endangered marine mammal, reptile, and bird species plus a few marine fish, mollusks, and corals. The structure of CITES limits its usefulness for protecting many marine species, but this could change with reinterpretation of the treaty or changes in the official status of marine species that are overexploited.[7] Greater restrictions on international trade would be helpful, as would economic incentives in the form of loans to promote alternate food sources and the establishment of enforceable protective programs.

In offshore and international waters, regulation is more difficult and is not prevalent. International agreements are needed and enforcement must be effective. Regulations should include more than just catch limits on target species. There needs to be some regulation on nontarget species, including fish species as well as mammals and turtles. There should also be off-limit designations for areas that are determined to be critical reproductive areas for commercial fisheries. Also needed are regulations on fishing technology. For example, technologies that flagrantly overharvest all species they are capable of catching, such as drift nets, should be banned by international agreement. In fact, the U.N. General Assembly recently passed a resolution calling for a worldwide ban on drift nets by June 30, 1992. Unfortunately, it is nonbinding and will be impossible to enforce, but it reflects an international climate favorable to more forceful agreements or conventions, such as a ban on drift nets in the South Pacific.

The difficulty of drafting, ratifying, implementing, and enforcing effective international agreements on the harvesting of living marine resources is exemplified by the International Whaling Commission. If it was so difficult to protect marine mammals, which are widely recognized as threatened and endangered, how much more difficult it will be to establish effective agreements for the protection of species whose status is less well defined.

Marine Protected Areas

While it has been stressed that marine protected areas alone will not save biological diversity in the oceans, it is also true that they may be very effective in conserving certain types of habitats and certain types of biological communities. For example, some coral reefs may be very well served by this method of protection. In areas where the greatest threat to the reefs is from overharvesting by tourists and commercial interests, establishing a marine park or reserve that prohibits or severely restricts harvesting could be effective. Of course, when the reef is near a coast where there is significant pollution, e.g., from deforestation, the protected area status may not help.

Reefs are particularly well suited to protected area status because they are physically defined areas harboring a characteristic diversity of species. These species tend to be either fixed in place or territorial and they therefore don't stray far from "home." Other benthic communities may also receive adequate protection from protected area status, but pelagic communities are less amenable to such methods. The effectiveness of the protection afforded animals and plants that remain within the sanctuary boundaries will depend largely on: a) the size of the area protected, b) what activities are restricted within the boundaries, and c) whether the protected area designation restricts polluting activities that occur outside the boundaries but threaten life within them. These are legally complicated issues that must be resolved on both national and international levels.

Importance of NGOs

As described above, there is a broad spectrum of research, legislation, regulation, and management practice that should be implemented, along with some well-placed economic incentives. For marine ecosystems, we are only at the stage of determining what is needed; the next stage will require a strong and widespread commitment to protecting living marine resources and marine biological diversity. The necessary initiatives will require large expenditures by governments and industries.

Also important, however, are persistent efforts by national and international environmental groups (often referred to as nongovernmental organizations, or NGOs) to press for the initiation of these programs. The major roles of these groups, in connection with marine biological diversity, will be to: (a) educate the public and government agencies as to the importance of protecting diversity, (b) analyze current policies with respect to diversity, (c) play an advocacy role in achieving important protective legislation and international conventions, and (d) participate in the oversight of implementation of protective laws and agreement. Since some of these endeavors involve compiling available information into suitable formats, these groups will also be able to identify some of the research that has to be done.

The need for educational programs is evidenced by the changes in awareness brought about by programs to educate the public on the crisis of the tropical rain forests. Among the people that should be the targets of educational programs on marine biological diversity are: those working on biological diversity issues; users of the oceans, such as fishermen, coastal residents, and tourists; and schoolchildren and the interested public.

Many NGOs are capable of providing in-depth analyses of national and international policy toward the goal of convincing the governments of the world to change those policies where that is deemed necessary. There are several policy issues pertinent to the preservation of marine biological diversity. A few that deserve careful review at both national and international levels are: (a) general policy on biological diversity, (b) policies on regulating fisheries and aquaculture, (c) policies on regulating ocean pollution from land-based sources, and (d) policies on coastal development.

There are several ways in which NGOs can play an important advocacy role with respect to marine biological diversity. They can work to persuade government officials to change national policies at the legislative level (U.S. Congress and other national parliaments) and at the administrative level (presidents or prime ministers and their cabinets), and they can influence international agreements by participating in formal meetings convened to consider proposed conventions and protocols. Other institutions where NGOs can have an effect in changing policy include the international lending agencies, such as the multilateral development banks. Environmental policy in these organizations has a tremendous effect on whether development in third world countries is sustainable and environmentally sound or not. Future economic development in the wealthy countries (i.e., the lenders) must be controlled by strong national environmental laws and international environmental treaties. It is critical that these be instated in tandem with strong environmental criteria for lending institutions, since in order to be effective the rules must apply to everyone.

The role of oversight is as important as advocacy. Once policies, laws, and international agreements are enacted, the implementation and enforcement must be timely and must stay on the right track. This does not happen automatically. It is important that the resulting regulations, criteria, management plans, monitoring, and enforcement actions be consistent with the spirit and goals of the laws and treaties. The process generally provides opportunity for public scrutiny and comment, and it usually falls upon NGOs to provide the public oversight, since technical expertise and familiarity with the laws are essential in evaluating the process. This oversight may involve a number of activities, including the analysis of environmental impact assessments, testifying at public hearings and congressional hearings, and, ultimately, litigation if the responsible government agencies are not properly implementing and enforcing the laws and treaties.

Grassroots groups, which are local citizens' groups generally involved in specific environmental problems affecting their "neighborhoods," also have an important role with respect to marine biological diversity. Coastal groups are able to identify areas where serious degradation of marine ecosystems is not being adequately addressed by responsible government agencies. They can often apply effective pressure on local and federal government officials to better enforce existing regulations and to support enactment of stronger laws and regulations. Perhaps the most important role of grassroots groups is to work with and garner the support of users of marine ecosystems, such as fishermen and tourists. When these people understand the environmental hazards and help forge a solution, damaged ecosystems can be put on the road to recovery more quickly.

Can We Commit Ourselves to Protecting Biodiversity?

The human species is unique on earth. We are unique in our ability to develop the world around us and to manage natural resources to suit our purposes. That means we can overdevelop the biosphere beyond its capacity to sustain itself or we can measure and understand the consequences of our actions and avoid the disaster if we choose. One potential major consequence of our actions is the loss of biological diversity on earth.

It is time for humans to strike a bargain with the earth. We must agree to provide for the maintenance of biological diversity and the sustained productivity of living systems. In exchange, we receive a reasonable portion of earth's bounty and a life-supporting global environment. Without the bargain, our own future is as dismal as that of the rest of the living world.

The authors hope that the foregoing discussion of biological diversity in marine environments has made it clear that any plans for maintaining the earth's living systems cannot neglect the oceans. Whether negotiating global agreements on the conservation of biological diversity or developing local and regional coastal zone management plans, the importance and diversity of biological functions in marine ecosystems should be recognized and factored into the bargain we humans strike with the natural world.

Glossary

algae: the simple photosynthetic plants (unicellular or multicellular, but not having specialized organs such as leaves, stems, and roots) found in the sea as well as in freshwater systems.

Microalgae include the microscopic (single-celled or colonies of cells) plants of the sea that are found wherever there is enough light for photosynthesis. They include a high diversity at higher taxonomic levels, with classes representing a broad spectrum of genetic information. The classes are: green, yellow-green, yellow-brown, golden-brown, brown, red, and blue-green algae.

Macroalgae, or *seaweeds,* include the larger plants of the sea that grow attached to the bottom from the high tide level on the shore down as deep as sufficient sunlight for photosyntheses penetrates. There are three classes of macroalgae: green, brown, and red algae.

anaerobic: living without oxygen. This is a term used to describe organisms living in oxygen-free environments.

anoxia: environmental condition in which there is no free dissolved oxygen.

anthropogenic: created or accomplished by humans.

benthic: upon or attached to the sea bottom.

biogeochemical cycles: the flow of elements through the earth's ecosystems, including various living and nonliving components of the environment. Physical and biological processes move elements from land to sea to atmosphere and back to land again, with the potential for some sequestering on land or in the ocean. Biogeochemical cycles important to the maintenance of the earth's atmosphere are the carbon, oxygen, hydrogen, and nitrogen cycles. Other major cycles include phosphorus, sulfur, calcium, sodium, and chlorine.

biogeography: the scientific study of the geographic distribution of organisms.

biome: a grouping of all communities or ecosystems worldwide having a similar biotic community and occurring in broadly similar environments.

139

biosphere: the part of the earth system (air/water/rock) that supports life.

biota: all the organisms found in a specified area.

biotic: pertaining to life or living organisms.

carnivore: an animal that eats other living animals.

chromosome: the molecular unit of inheritance in living organisms. Each species has a characteristic number of chromosomes (e.g., 48 for humans), identical in each cell of an organism, that carry all the genetic information for that organism.

community, or **biological community:** the full complement of species that inhabit a specified habitat.

decomposition: the breakdown of dead organic material into simpler molecules (bacteria and fungi are the primary *decomposers* of the biosphere).

demersal: living near but not upon the bottom (e.g., demersal fish).

detritivore: an animal that eats detritus (dead organic material). A large detritivore, while feeding on detritus, may also ingest small living organisms (such as bacteria) associated with it.

ecosystem: a biological community and its physical environment.

endangered species: a term that legally means any species determined to be reduced to a global population level that is close to or beneath the sustainable level for that species (the level is determined by scientists, who make the best guess they can based on the information they have about the species). Since the species is in imminent danger of going extinct, it is afforded protection under laws recognized by many nations.

endemic (endemism): found only in one location or habitat. This term is usually applied to species.

eutrophication: the process of nutrient enrichment of an aquatic ecosystem leading to increased biologic production. As eutrophication proceeds, there are a number of consequences, including excess production, increased decay, reduced oxygen, and decreased biological diversity.

evapotranspire: the process by which a land plant that has roots, stems, and leaves moves water into the roots, up the stem (or trunk), and into the leaves, from which it then evaporates into the air.

fauna: all the animals in a specified habitat.

filter feeding: a method of feeding (found only in the water) by which an animal moves water past some structure it has that is capable of filtering out particles within a particular size range. Those particles, which include the animal's preferred food organisms, are then ingested. Barnacles, which filter very small plankton from the water, and baleen whales, which filter larger plankton (krill), are two common examples of filter feeders.

food web, or **food chain:** the diagrammatic representation of who feeds on whom in a biotic community. In general, photosynthetic plants are the starting point or the primary producers, and a network of animals feeding on plants and animals feeding on animals describes the transfer of energy through the entire community (or ecosystem). Food webs may specify particular species, but often they merely describe groups of organisms that utilize a similar source of food.

gene: the basic unit of heredity. Each species has a characteristic number of chromosomes, each of which contains a series of genes that are the chemical basis of inherited traits.

gene pool: all the genes in a population.

habitat: the environment in which an organism, species, or community lives.

herbivore: an animal that eats plants.

invertebrates: all kinds of animals lacking a backbone, from protozoans to insects and starfish.

larva (pl. **larvae**): in certain animal species, an immature stage that is radically different in form from the adult, characteristic of many marine invertebrates.

mollusk: a member of a large phylum of invertebrate animals (Mollusca) characterized by soft, unsegmented bodies and usually having a calcareous shell.

niche: the range of each environmental variable, such as temperature, salinity, nutrients, and food items, within which a species can exist and reproduce. The preferred niche is the one in which the species performs best in the absence of competition or interference from extraneous factors. The realized niche is the one in which it actually comes to live in a particular environment.

organism: an individual living creature.

pelagic: living in the ocean water column.

photosynthesis: a photochemical reaction in cells with chlorophyll, in which solar energy, carbon dioxide, and water are converted into glucose and eventually into more complicated molecules. The energy stored in photosynthesizing organisms becomes available to animals that eat and digest the plant material or to decay organisms, which break down dead organisms.

phytoplankton: microscopic algae that drift in sunlit surface waters.

pollution: the contamination of a natural ecosystem by wastes from human activities. The contaminants may be nutrients that initially stimulate growth of primary producers, or they may be chronic toxins.

population: a set of organisms belonging to the same species and occupying a clearly delimited space at the same time.

productivity: the rate of biological production in an ecosystem. Primary productivity is the rate of transformation of solar or chemical energy to living material (biomass). Production refers to the amount of material produced, whereas productivity is a rate of production.

protista: the kingdom Protista, which includes unicellular organisms with distinct internal cellular structure, including protozoa and unicellular algae. (Bacteria and blue-green algae are more primitive single-celled organisms, without an internal cellular structure, and they belong to the kingdom Monera.)

protozoan (pl. **protozoa**): a group of single-celled protista that require an organic food source; i.e., they cannot synthesize food from light or chemical energy. Therefore, they are often considered more animallike than plantlike.

species: a population or collection of populations of closely related and similar organisms capable of interbreeding freely with one another but not with members of other species under natural conditions.

species diversity: the number of species in a region, but sometimes defined as a function of the pattern of distribution as well as the abundance of species.

species richness: the number of species in a region. *Species diversity* is often used interchangeably with this term.

spore: a nonsexual reproductive cell capable of developing into an adult without fusion with another cell. It may contain the exact genetic material of the parent organism and grow into an identical offspring. Or it may contain half the number of chromosomes of the parent (similar to a sexual reproductive cell) and grow into an alternative form of that species; the alternate form then produces sexual reproductive cells (*gametes*) that fuse to produce an individual of the original spore-producing type. This latter type of life cycle exhibits *alternation of generations,* which is typical of many plants, including algae.

stress: an environmental factor that has a negative effect on an organism, a species, or a community.

systematics: the study of the evolutionary and genetic relationships among organisms and the development of a natural classification system based on those relationships.

taxon (pl. **taxa**): any group of organisms representing a particular unit of classification. In ascending order of inclusiveness, the taxa are: species, genus (pl. genera), family, order, class, phylum, kingdom. (Note: A single species will also belong to a particular genus, family, etc.)

taxonomy: the naming and assignment of organisms to taxa.

terrestrial: of or living on land.

trophic level: position of an organism in the food chain, determined by the number of transfers of energy that occur between the nonliving energy source and that position. Trophic levels include producers (photosynthesizers and chemosynthesizers that convert light or chemical energy into living material) and several levels of consumers (animals eating plants, animals eating animals eating animals . . .).

vascular plants: plants with specialized tissues for transporting water, minerals, and other nutrients from roots to leaves, and with tissues for transporting synthesized sugars from leaves to stems and roots. Sea grasses are vascular plants, but otherwise all marine plants are nonvascular. This is not surprising, since plants growing submerged in water do not have much need for special mechanisms to transport water with its dissolved nutrients to different portions of the plant body.

zonation: community stratification. In intertidal areas, there is a tendency for biotic assemblages to distribute themselves in horizontal strata, each assemblage located within the range of tidal height in which they most successfully balance productivity in response to light and length of exposure to the air against competition from other species.

zooplankton: small, sometimes microscopic animals that drift in the ocean. Zooplankton include protozoa, crustaceans, jellyfish, and other invertebrates that drift at various depths in the water column.

Notes

Introduction

1. R. C. Lewontin. 1990. Fallen angels. *The New York Review of Books* 37(10):3–7.

Chapter 1

1. G. C. Ray. 1988. Ecological diversity in coastal zones and oceans. Chapter 4 in *Biodiversity*, E. O. Wilson (Ed.), National Academy Press, Washington, D.C., pp. 36–50.
2. U.S. Congress, Office of Technology Assessment. 1987. *Technologies to maintain biological diversity*. OTA-F-330. U.S. Government Printing Office, Washington, D.C.
3. J. E. Lovelock. 1979. *Gaia: A new look at life on Earth*. Oxford University Press, New York.
4. D. Lindley. 1988. Is the Earth alive or dead? *Nature* 332:483–484.
5. J. Gribbin. 1988. The oceanic key to climatic change. *New Scientist* (May 19):32–33; W. M. Post, et al. 1990. The global carbon cycle. *American Scientist* 78:310–326.
6. World Resources Institute, International Institute for Environment and Development. 1987. *World resources 1987*. Basic Books, New York.
7. N. Myers. 1979. *The sinking ark*. Pergamon Press, Oxford, pp. 79–81.
8. R. R. Colwell. 1983. Biotechnology in the marine sciences. *Science* 222:23.
9. R. Tiner. 1984. *Wetlands of the United States: current status and recent trends*. U.S. Department of Interior, Washington, D.C.
10. M. D. Fortes. 1988. Mangrove and seagrass beds of East Asia: habitats under stress. *Ambio* 17:207–213.
11. J. W. McManus. 1988. Coral reefs of the ASEAN region: status and management. *Ambio* 17:189–193.
12. The Coastal Ocean Space Utilization Symposium was held in May 1989. Proceedings are being published by Elsevier Science Publishing Company, Inc., New York.
13. D. Fisher, et al. 1988. *Polluted Coastal Waters: the role of acid rain*. Environmental Defense Fund, New York.

14. GESAMP (Joint Group of Experts on the Scientific Aspects of Marine Pollution). 1990. *The State of the Marine Environment.* UNEP Regional Seas Reports and Studies No. 115. UNEP, Nairobi.
15. A. M. Manville. 1988. Tracking plastic in the Pacific. *Defenders* (Nov./Dec.):10–15.
16. L. R. Brown, et al. 1985. Maintaining world fisheries. Chapter 4 in *State of the world 1985,* L. R. Brown et al. W. W. Norton and Company, New York, pp. 74–95.
17. Ibid.
18. W. V. Reid and K. R. Miller. 1989. *Keeping options alive: The scientific basis for conserving biodiversity.* World Resources Institute, Washington, D.C., p. 39.
19. World Commission on Environment and Development. 1987. *Our common future.* Oxford University Press, New York, p. 264.

Chapter 2

1. S. L. Pimm. 1984. The complexity and stability of ecosystems. *Nature* 307:321–326.
2. R. M. May. 1988. How many species are there on Earth? *Science* 241:1441–1448.
3. R. S. Burton. 1983. Protein polymorphisms and genetic differentiation of marine invertebrate populations. *Marine Biology Letters* 4:193–206.
4. J. C. Gallagher. 1980. Population genetics of *Skeletonema costatum* (Bacillariophyceae) in Narragansett Bay. *Journal of Phycology* 16:464–474.
5. P. J. Smith and Y. Fugio. 1982. Genetic variation in marine teleosts: High variability in habitat specialists and low variability in habitat generalists. *Marine Biology* 69:7–20.
6. L. E. Brand. 1981. Genetic variability in reproduction rates in marine phytoplankton populations. *Estuaries* 35:1117–1127.
7. J. P. Grassle and J. F. Grassle. 1976. Sibling species in the marine pollution indicator *Capitella* (Polychaeta). *Science* 192:567–569.
8. U. Gyllensten and N. Ryman. 1985. Pollution biomonitoring programs and the genetic structure of indicator species. *Ambio* 14:29–31.

9. G. C. Ray. 1988. Ecological diversity in coastal zones and oceans. Chapter 4 in *Biodiversity*, E. O. Wilson (Ed.), National Academy Press, Washington, D.C., pp. 36–50. Also: May, 1988.

10. May, 1988.

11. Ray, 1988.

12. May, 1988.

13. A. R. Emery. 1978. The basis of fish community structure: Marine and freshwater comparisons. *Environmental Biology and Fisheries* 3:33–47.

14. J. H. Steele. 1985. A comparison of terrestrial and marine ecological systems. *Nature* 313:355–358.

15. E. C. Pielou. 1979. *Biogeography*. Wiley-Interscience, New York, p. 15.

16. W. Fenical. 1982. Natural products chemistry in the marine environment. *Science* 215:923–928.

17. D. E. Morse and A. N. C. Morse. 1988. Chemical signals and molecular mechanisms: Learning from larvae. *Oceanus* 31:37–43.

18. G. D. Ruggieri. 1976. Drugs from the sea. *Science* 194:491–497.

19. R. R. Colwell. 1983. Biotechnology in the marine sciences. *Science* 222:19–24.

20. F. G. Stehli, R. G. Douglas, and N. D. Newell. 1969. *Science* 164:947–949.

21. E. R. Pianka. 1988. *Evolutionary ecology*. Harper and Row, New York. pp. 309–314.

22. J. F. Grassle. 1989. Species diversity in deep-sea communities. *Trends in Ecology and Evolution* 4:12–15.

23. Ray, 1988.

24. G. J. Vermeij. 1978. *Biogeography and adaptation: Patterns of marine life*. Harvard University Press, Cambridge, Mass., p. 161.

25. E. B. Sherr. 1989. And now, small is plentiful. *Nature* 340:429.

26. P. K. Dayton. 1984. Processes structuring some marine communities: Are they general? Chapter 12 in *Ecological Communities: Conceptual Issues and Evidence*, D. R. Strong, Jr. (Ed.), Princeton University Press, Princeton, New Jersey, pp. 181–197.

27. Ibid.

28. B. P. Hayden, G. C. Ray, and R. Dolan. 1984. Classification of coastal and marine environments. *Environmental Conservation* 11:199–207.

Chapter 3

1. A. J. Underwood and E. J. Denley. 1984. Paradigms, explanations, and generalizations in models for the structure of intertidal communities on rocky shores. Chapter 11 in *Ecological Communities: Conceptual Issues and the Evidence,* D. R. Strong, Jr. (Ed.), et al., pp. 151–180.
2. R. T. Paine. 1966. Food web complexity and species diversity. *American Naturalist* 100:65–75. And: R. T. Paine. 1984. Some approaches to modeling multispecies systems. In *Exploitation of Marine Communities,* R. M. May (Ed.), Springer-Verlag, New York, pp. 191–207.
3. R. T. Paine, J. C. Castillo, and J. Cancino. 1985. Perturbation and recovery patterns of starfish-dominated intertidal assemblages in Chile, New Zealand and Washington State. *American Naturalist* 125:679–691.
4. P. K. Dayton. 1984. Processes structuring some marine communities: Are they general? Chapter 12 in *Ecological Communities: Conceptual Issues and Evidence,* D. R. Strong, Jr. (Ed.) et al. Princeton University Press, Princeton, New Jersey, pp. 181–197. And: J. Roughgarden, S. Gaines, and H. Possingham. 1988. Recruitment dynamics in complex life cycles. *Science* 241:1460–1466.
5. T. T. Spight. 1977. Diversity of shallow-water gastropod communities on temperate and tropical beaches. *American Naturalist* 982:1077–1097.
6. S. Pain. 1988. No escape from the global greenhouse. *New Scientist* (November 12): p. 38.
7. J. Lewin. 1978. The world of the razor-clam beach. *Pacific Search* (April):12–13.
8. D. F. Boesch. 1974. Diversity, stability and response to human disturbance in estuarine ecosystems. *Proceedings of the First International Congress of Ecology, The Hague, The Netherlands, September 8–14, 1974.* Pudoc, Wagenigen, The Netherlands, pp. 109–114.
9. D. S. McLusky. 1981. *The estuarine ecosystem.* John Wiley and Sons, New York.

10. R. G. Wiegert and L. R. Pomeroy. 1981. The salt-marsh ecosystem: A synthesis. Chapter 10 in *The Ecology of a Salt Marsh*, R. Pomeroy and R. G. Wiegert (Eds.), Springer-Verlag, New York, pp. 220–221.
11. Pain, 1988.
12. J. W. Wells. 1957. Coral reefs. Chapter 20 in *Treatise on Marine Ecology and Paleoecology*, J. W. Hedgpeth (Ed.), Memoir 67, vol. 1, Geological Society of America, New York, p. 601.
13. J. H. Connell. 1978. Diversity in tropical rain forests and coral reefs. *Science* 199:1302–1310.
14. M. A. Huston. 1985. Patterns of species diversity on coral reefs. *Annual Review of Ecology and Systematics* 16:149–177.
15. P. F. Sale. 1980. The ecology of fishes on coral reefs. *Oceanography and Marine Biology, Annual Review* 18:367–421.
16. F. G. Stehli and J. W. Wells. 1971. Diversity and patterns in hermatypic corals. *Systematic Zoology* 20:115–126.
17. C. Birkeland. 1990. Geographic comparisons of coral-reef community processes. In *Proceedings of the 6th International Coral Reef Symposium 1988*.
18. J. Ogden. 1989. Marine biological diversity: a strategy for action. *Reef Encounter* 6:5.
19. D. M. Alongi. 1990. The ecology of tropical soft-bottom benthic ecosystems. *Oceanography and Marine Biology Annual Review* 28:381–496.
20. K. Sherman et al. 1988. The continental shelf ecosystem off the northeast coast of the United States. In *Ecosystems of the World, 27: Continental Shelves*, H. Postma and J. J. Zijlstra (Eds.), Elsevier, Amsterdam, pp. 279–337.
21. B. Rygg. 1985. Distribution of species along pollution-induced diversity gradients in benthic communities in Norwegian fjords. *Marine Pollution Bulletin* 16:469–474.
22. A. Bucklin. 1986. The genetic structure of zooplankton populations. In *Pelagic Biogeography: Proceedings of an International Conference, The Netherlands, 29 May–5 June, 1985*. UNESCO Technical Papers in Marine Science, No. 49, pp. 33–41.
23. K. Sherman et al. 1983. Coherence in zooplankton of a large northwest Atlantic ecosystem. *Fishery Bulletin* 81:855–862.
24. M. P. Sissenwine. 1986. Perturbation of a predator-controlled continental shelf ecosystem. Chapter 5 in *Variability and Man-*

agement of Large Marine Ecosystems, K. Sherman and L. Alexander (Eds.), AAAS Selected Symposium 99:55–85.

Chapter 4

1. J. J. Stegman et al. 1986. Monooxygenase induction and chlorobiphenyls in the deep-sea fish *Coryphaenoides armatus. Science* 231:1287–1289.
2. H. L. Sanders. 1968. Marine benthic diversity: A comparative study. *American Naturalist* 102:243–282.
3. P. A. Jumars. 1976. Deep-sea species diversity: Does it have a characteristic scale? *Journal of Marine Research* 34:217–246.
4. L. G. Abele and K. Walters. 1979. The stability-time hypothesis: Reevaluation of the data. *American Naturalist* 114:559–568.
5. M. A. Rex. 1981. Community structure in deep sea benthos. *Annual Review of Ecology and Systematics* 12:331–353.
6. J. F. Grassle. 1989. Species diversity in deep-sea communities. *Trends in Ecology and Evolution* 4:12–15. And: J. F. Grassle and N. J. Maciolek. In preparation. Deep-sea species richness: Regional and local diversity estimates from quantitative bottom samples.
7. Grassle, 1989.
8. Jumars, 1976.
9. P. K. Dayton and R. R. Hessler. 1972. Role of biological disturbance in maintaining diversity in the deep sea. *Deep-Sea Research* 19:199–208.
10. A. Aarkrog et al. 1987. Technetium-99 and Cesium-134 as long distance tracers in Arctic waters. *Estuarine, Coastal and Shelf Science* 24:637–647. And: H. D. Livingston et al. 1984. Vertical profile of artificial radionuclide concentrations in the central Arctic Ocean. *Geochimica and Cosmochimica Acta* 48:2195–2203.
11. R. Monastersky. 1988. Plentiful plankton noticed at last. *Science News* 134:68.
12. Y. A. Rudyakov. 1987. Ecosystems of coastal waters as a component of the biological structure of the ocean. *Oceanology* 27:479–481.
13. R. Margalef. 1968. *Perspectives in Ecological Theory.* University of Chicago Press, p. 52.
14. R. Margalef. 1968.

15. G. D. Grice and A. D. Hart. 1962. The abundance, seasonal occurrence and distribution of the epizooplankton between New York and Bermuda. *Ecological Monographs* 32:287–309.
16. Ibid.
17. S. J. Giovanni et al. 1990. Genetic diversity in Sargasso Sea bacterioplankton. *Nature* 345:60.
18. J. A. McGowan. 1986. The biogeography of pelagic ecosystems. In *Pelagic Biogeography: Proceedings of an International Conference, The Netherlands, 29 May-5 June, 1985.* UNESCO Technical Papers in Marine Science, No. 49, pp. 191–200.
19. J. A. McGowan and P. W. Walker. 1979. Structure in the copepod community of the North Pacific Central gyre. *Ecological Monographs* 49:195–226.
20. M. A. Barnett. 1983. Species structure and temporal stability of mesopelagic fish assemblages in the central gyres of the North and South Pacific Ocean. *Marine Biology* 74:245–256.
21. E. L. Venrick. 1982. Phytoplankton in an oligotrophic ocean: Observations and questions. *Ecological Monographs* 52:129–154.
22. B. S. Manheim, Jr. 1988. *On thin ice: The failure of the National Science Foundation to protect Antarctica.* Environmental Defense Fund, August 1988.
23. National Science Board. 1988. *The role of the National Science Foundation in polar regions.* NSB-87-128. National Science Foundation, Washington, D.C.

Chapter 5

1. R. Margalef. 1968. *Perspectives in ecological theory.* University of Chicago Press.
2. R. V. Salm with J. R. Clark. 1984. *Marine and coastal protected areas: A guide for planners and managers.* IUCN, Gland, Switzerland, p. 4.
3. For a discussion of some of these differing views, see Volume 33 of *Oceanus* Summer, 1990.
4. J. H. Steele et al. 1989. *Comparison of terrestrial and marine ecological systems: Report of a workshop held in Santa Fe, New Mexico.*
5. J. A. McNeely. 1988. *Economics and Biological Diversity.* IUCN, Gland, Switzerland.

6. R. Costanza. 1989. What is Ecological Economics? *Ecological Economics* 1:1–7.

Chapter 6

1. U.S. Congress, House, Committee on Merchant Marine Fisheries, 1988. *Report on the National Marine Sanctuaries Authorization Act of 1988*. Report 100-739, Part 1 to accompany H.R. 4208. 100th Congress, 2nd Session.
2. Council on Environmental Quality, 1988. *Environmental quality—1986: 17th annual report of the Council on Environmental Quality*. U.S. Government Printing Office, Washington, D.C.
3. 15 CFR Part 921, Section 921.2, 53 Federal Register 43821, October 28, 1988.
4. The U.S. Fish and Wildlife Service has responsibility for polar bears, sea otters, manatees, and sea turtles when they are nesting on land; the National Marine Fisheries Service has responsibility for all other marine species.
5. Endangered Species Act. 16. U.S.C. 1532(s)(A).
6. From list compiled by Defenders of Wildlife, 1988.
7. The manatee, Hawaiian monk seal, and hawksbill and leatherback sea turtles.
8. Maximum net productivity is the greatest net annual increment in population numbers or biomass resulting from additions to the population due to reproduction, growth, less losses due to natural mortality. (See 41 Federal Register 55536, December 21, 1976.)
9. The northern fur seal, Hawaiian monk seal, bowhead whale, and all endangered and threatened marine mammals are designated as depleted under MMPA.
10. Until the inclusion of tuna in 1990, it remained the only fishery resource not covered under the act. Because of the highly migratory nature of tuna, the species is managed by international agreement.
11. The term *exclusive economic zone* was adopted by the United Nations during the UN Conference on the Law of the Sea and is embodied in the 1982 Law of the Sea Convention.
12. U.S. Congress, Office of Technology Assessment, 1987. *Wastes in Marine Environments*, OTA-O-334. U.S. Government Printing Office, Washington, D.C., p. 154.

13. "Dumping" under the Ocean Dumping Act refers to the disposal of any material into the ocean other than by the discharge of effluent from outfalls (which is regulated under the Clean Water Act) or the discharge of material from vessels during the normal operation of such vessel.
14. From Section 102(a) of Public Law 92-532 (The Marine Protection, Research, and Sanctuaries Act of 1972).
15. The White House, Office of the Press Secretary, 1990. *Fact Sheet: Presidential Decisions Concerning Oil and Gas Development on the Outer Continental Shelf.* June 26, 1990.

Chapter 7

1. Areas covered by the UNEP Regional Seas Programme include the Mediterranean Sea, Arabian Gulf, West and Central Africa, the Southeast Pacific, the Red Sea–Gulf of Aden, the Caribbean Sea, East Asia Seas, South Pacific, and Southwest Atlantic, East Africa, and South Asian Sea.
2. UNEP. 1982. *Achievements and planned development of UNEP's Regional Seas Programme and comparable programmes sponsored by other bodies.* UNEP Regional Seas Reports and Studies No. 1, UNEP, Nairobi.
3. IUCN. 1984. *International Union for the Conservation of Nature and Natural Resources triennial report, 1982–84.* Gland, Switzerland.
4. IUCN. 1980. *World conservation strategy.* Gland, Switzerland.
5. IUCN. 1987. *Elements of the IUCN Coastal and Marine Conservation Programme.* Gland, Switzerland.
6. J. A. Heap and M. W. Holdgate. 1986. The Antarctic treaty system as an environmental mechanism—An approach to environmental issues. In *Antarctic Treaty System: An Assessment,* National Research Council, National Academy Press, Washington, D.C., pp. 195–210.
7. U.S. Agency for International Development. 1985. *U.S. strategy on the conservation of biological diversity: An interagency task force report to Congress.* U.S. AID, Washington, D.C.
8. U.S. AID. 1988. *Progress in conserving tropical forests and biological diversity in developing countries: The 1987 annual report to Congress on the implementation of Sections 118 and 119 of the Foreign Assistance Act, as amended.* U.S. AID, Washington, D.C.

9. B. Hulshoff and W. P. Gregg. 1985. Biosphere reserves: Demonstrating the value of conservation in sustaining society. *Parks* 10:2–5.

10. World Resources Institute, International Institute for Environment and Development. 1988. *World resources 1988–89*. Basic Books, New York. pp. 143–161.

11. T. Agardy. 1988. *A status report on the workings of the joint U.S./Canada ad hoc selection panel for Acadian boreal biosphere reserve nomination*. Prepared for the U.S. MAB Directorate on Biosphere Reserves. U.S. Department of State, Washington, D.C.

Chapter 8

1. The gliding submersible, "Deep Flight," is being developed by Deep Ocean Engineering in San Leandro, California.

2. World Commission on Environment and Development. 1987. *Our common future*. Oxford University Press, New York, p. 264.

3. J. A. McNeely. 1988. *Economics and biological diversity*. IUCN, Gland, Switzerland.

4. S. Olsen, L. Z. Hale, R. DuBois, D. Robadue, and G. Foer. 1989. *Integrated resources management for coastal environments in the Asia Near East Region*. International Coastal Resources Management Project, University of Rhode Island.

5. Ibid.

6. McNeely, 1988.

7. S. M. Wells and J. G. Barzdo. (in press). International trade in marine species—is CITES a useful control mechanism? *Journal of Coastal Management*.

Bibliography

Aarkrog, A., S. Boelskifte, H. Dahlgaard, S. Duniec, L. Hallstadius, E. Holm, and J. N. Smith. 1987. Technetium-99 and Cesium-134 as long distance tracers in Arctic waters. *Estuarine, Coastal and Shelf Science* 24:637–647.

Abele, L. G., and K. Walters. 1979. The stability-time hypothesis: Reevaluation of the data. *American Naturalist* 114:559–568.

Agardy, T. 1988. *A status report on the workings of the joint U.S./ Canada ad hoc selection panel for Acadian boreal biosphere reserve nomination.* Prepared for the U.S. MAB Directorate on Biosphere Reserves, U.S. Department of State, Washington, D.C.

Alongi, D. M. 1990. The ecology of tropical soft-bottom benthic ecosystems. *Oceanography and Marine Biology Annual Review* 28:381–496.

Ayala, F. J., and J. W. Valentine. 1979. Genetic variability in the pelagic environment: A paradox. *Ecology* 60:24–29.

Bak, R. P. M. 1987. Effects of chronic oil pollution on a Caribbean coral reef. *Marine Pollution Bulletin* 18:534–539.

Barnett, M. A. 1983. Species structure and temporal stability of mesopelagic fish assemblages in the central gyres of the North and South Pacific Ocean. *Marine Biology* 74:245–256.

Belsky, M. 1986. Legal constraints and options for total ecosystem management of large marine ecosystems. Chapter 11 in *Variability and Management of Large Marine Ecosystems*, K. Sherman and L. Alexander (Eds.), AAAS Selected Symposium 99:241–262.

Ben-Eliahu, M. N., and U. N. Safriel. 1982. A comparison between species diversities of polychaetes from tropical and temperate structurally similar rocky intertidal habitats. *Journal of Biogeography* 9:371–390.

Ben-Eliahu, M. N., U. N. Safriel, and S. Ben-Tuvia. 1988. Environmental stability is low where polychaete species diversity is high: Quantifying tropical vs. temperate within-habitat features. *Oikos* 52:255–273.

155

Bibliography

Birkeland, C. 1990. Geographic comparisons of coral-reef community processes. In *Proceedings of the 6th International Coral Reef Symposium, 1988.* Queensland, Australia.

Boesch, D. F. 1974. Diversity, stability, and response to human disturbance in estuarine ecosystems. In *Structure, Functioning and Management of Ecosystems: Proceedings of the First International Congress of Ecology, The Hague, The Netherlands, September 8–14, 1974,* Pudoc, Wagenigen, The Netherlands, pp. 109–114.

Boyle, E. A. 1990. Quaternary deep water paleooceanography. *Science* 249:863–869.

Brand, L. E. 1981. Genetic variability in reproduction rates in marine phytoplankton populations. *Estuaries* 35:1117–1127.

Brown, L.R., W. U. Chandler, C. Flavin, C. Pollock, S. Postel, L. Starke, and E. C. Wolf. 1985. *State of the World 1985.* W. W. Norton and Company, New York.

Bucklin, A. 1986. The genetic structure of zooplankton populations. In *Pelagic Biogeography: Proceedings of an International Conference, The Netherlands, 29 May–5 June, 1985,* UNESCO Technical Papers in Marine Science, No. 49, pp. 33–41.

Burton, R. S. 1983. Protein polymorphisms and genetic differentiation of marine invertebrate populations. *Marine Biology Letters* 4:193–206.

Carlson, C. 1988. NEPA and the conservation of biological diversity. *Environmental Law* 19:15–36.

Colwell, R. R. 1983. Biotechnology in the marine sciences. *Science* 222:19–24.

Connell, J. H. 1978. Diversity in tropical rain forests and coral reefs. *Science* 199:1302–1310.

Connell, J. H. 1983. On the prevalence and relative importance of interspecific competition: Evidence from field experiments. *American Naturalist* 122:661–696.

Council on Environmental Quality. 1980. *Environmental quality— 1980: The eleventh annual report of the Council on Environmental Quality.* U.S. Government Printing Office, Washington, D.C.

Council on Environmental Quality. 1986. *Environmental quality— 1985: 16th annual report of the Council on Environmental Quality.* U.S. Government Printing Office, Washington, D.C.

Council on Environmental Quality. 1988. *Environmental quality— 1986: 17th annual report of the Council on Environmental Quality.* U.S. Government Printing Office, Washington, D.C.

Curtis, C. 1988. Environmental conditions and trends in marine and near-coastal waters. Testimony before the Subcommittee on Environmental Protection of the Senate Committee on Environment and Public Works, April 20.

Curtis, C. E. 1990. Protecting the oceans. *Oceanus* 3:19–22.

Dayton, P. K. 1984. Processes structuring some marine communities: Are they general? Chapter 12 in *Ecological Communities: Conceptual Issues and Evidence*, D. R. Strong, Jr. (Ed.), pp. 181–197.

Dayton, P. K., and R. R. Hessler. 1972. Role of biological disturbance in maintaining diversity in the deep sea. *Deep-Sea Research* 19:199–208.

Dunbar, M. J. 1968. *Ecological development in polar regions: A Study in evolution.* Prentice-Hall, Inc., Englewood Cliffs, New Jersey.

Emery, A. R. 1978. The basis of fish community structure: Marine and freshwater comparisons. *Environmental Biology and Fisheries* 3:33–47.

Fenical, W. 1982. Natural products chemistry in the marine environment. *Science* 215:923–928.

Fisher, D., J. Ceruso, T. Mathew and M. Oppenheimer. 1988. *Polluted Coastal Waters: the role of acid rain.* Environmental Defense Fund, N.Y. 102 pp.

Foster, N., and M. H. Lemay (Eds.). 1988. *Managing marine protected areas: An action plan.* Department of State Publication number 9673. U.S. Department of State, Man and the Biosphere Program, Washington, D.C.

Gallagher, J. C. 1980. Population genetics of *Skeletonema costatum* (Bacillariophyceae) in Narragansett Bay. *Journal of Phycology* 16:464–474.

GESAMP (Joint Group of Experts on the Scientific Aspects of Marine Pollution). 1990. The state of the marine environment. UNEP Regional Seas Reports and Studies No. 115. UNEP, Nairobi. 111 pp.

Giovannoni, S. J., T. B. Britschgi, C. L. Moyer, and K. G. Field. 1990. Genetic diversity in Sargasso Sea bacterioplankton. *Nature* 345:60–61.

Gittings, S. R., and T. J. Bright. 1988. The *M/V Wellwood* grounding: A sanctuary case study. The science. *Oceanus* 31:36–41.

Grassle, J. F. 1989. Species diversity in deep-sea communities. *Trends in Ecology and Evolution* 4:12–15.

Grassle, J. F., and N. J. Maciolek. In preparation. Deep-sea species richness: Regional and local diversity estimates from quantitative bottom samples.

Grassle, J. F., and H. L. Sanders. 1973. Life histories and the role of disturbance. *Deep-Sea Research* 20:643–659.

Grassle, J. F., N. J. Maciolek, and J. A. Blake. In press. Are deep-sea communities resilient? Chapter 17 in *The Earth in Transition: Patterns and Processes of Biotic Impoverishment*, G. Woodwell (Ed.), Proceedings of a conference held at the Woods Hole Research Center, Woods Hole, Mass., October 1989.

Grassle, J. P., and J. F. Grassle. 1976. Sibling species in the marine pollution indicator *Capitella* (Polychaeta). *Science* 192: 567–569.

Gribbin, J. 1988. The oceanic key to climatic change. *New Scientist* (May 19):32–33.

Grice, G. D., and A. D. Hart. 1962. The abundance, seasonal occurrence and distribution of the epizooplankton between New York and Bermuda. *Ecological Monographs* 32:287–309.

Gyllensten, U., and N. Ryman. 1985. Pollution biomonitoring programs and the genetic structure of indicator species. *Ambio* 14:29–31.

Hardy, J. T., and C. W. Apts. 1989. Photosynthetic carbon reduction: High rates in the sea-surface microlayer. *Marine Biology* 101:411–417.

Hart, S. 1989. Food chains: The carbon link. *Science* 136:168–170.

Hayden, B. P., G. C. Ray, and R. Dolan. 1984. Classification of coastal and marine environments. *Environmental Conservation* 11:199–207.

Heap, J. A., and M. W. Holdgate. 1986. The Antarctic treaty system as an environmental mechanism—An approach to environmental issues. In *Antarctic Treaty System: An Assessment*, National Research Council, National Academy Press, Washington, D.C., pp. 195–210.

Hessler, R. R., and H. L. Sanders. 1967. Faunal diversity in the deep-sea. *Deep-Sea Research* 14:65–78.

Hulshoff, B., and W. P. Gregg. 1985. Biosphere reserves: Demonstrating the value of conservation in sustaining society. *Parks* 10:2–5.

Huston, M. A. 1985. Patterns of species diversity on coral reefs. *Annual Review of Ecology and Systematics* 16:149–177.

International Union for the Conservation of Nature and Natural Resources. 1980. *World conservation strategy*. IUCN, Gland, Switzerland.

International Union for the Conservation of Nature and Natural Resources. 1984. *International Union for the Conservation of Nature and Natural Resources triennial report, 1982–1984*. IUCN, Gland, Switzerland.

International Union for the Conservation of Nature and Natural Resources. 1987. *Elements of the IUCN coastal and marine conservation programme*. IUCN, Gland, Switzerland.

Jackson, J. B. C., and K. W. Kaufmann. 1987. *Diadema antillarum* was not a keystone predator in cryptic reef environments. *Science* 235:687–689.

Johannes, R. E. 1978. Traditional marine conservation methods in Oceania and their demise. *Annual Review of Ecology and Systematics* 9:349–364.

Joint Group of Experts on the Scientific Aspects of Marine Pollution (GESAMP). 1990. *The state of the marine environment*. UNEP Regional Seas Reports and Studies no. 115. United Nations Environmental Programme. Nairobi.

Jumars, P. A. 1976. Deep-sea species diversity: Does it have a characteristic scale? *Journal of Marine Research* 34:217–246.

Kerr, R. 1983. Are the ocean's deserts blooming? *Science* 220:397–398.

Kingston, P. F. 1987. Field effects of platform discharges on benthic macrofauna. *Philosophical Transactions of the Royal Society of London* B 316:545–565.

Leigh, E. G., Jr., R. T. Paine, J. F. Quinn, and T. H. Suchanek. 1984. Wave energy and intertidal productivity. *Proceedings of the National Academy of Sciences, U.S.A.* 84:1314–1318.

Lewin, J. 1978. The world of the razor-clam beach. *Pacific Search* (April):12–13.

Lindley, D. 1988. Is the Earth alive or dead? *Nature* 332:483–484.

Livingston, H. D., S. L. Kupferman, V. T. Bowen, and R. M. Moore. 1984. Vertical profile of artificial radionuclide concentrations in the central Arctic Ocean. *Geochimica and Cosmochimica Acta* 48:2195–2203.

Lovelock, J. E. 1979. *Gaia: A new look at life on Earth.* Oxford University Press, New York.

Lovelock, J. E. 1988. The earth as a living organism. Chapter 56 in *Biodiversity*, E. O. Wilson (Ed.), National Academy Press, Washington D.C., pp. 487–489.

Lugo, A. E., and S. C. Snedaker. 1974. The ecology of mangroves. *Annual Review of Ecology and Systematics* 5:39–64.

Lynch, M. P., and G. C. Ray. 1985. Diversity of marine/coastal ecosystems. Paper prepared under contract to the U.S. Congressional Office of Technology Assessment.

Manheim, B. S., Jr. 1988. *On thin ice: The failure of the National Science Foundation to protect Antarctica.* Environmental Defense Fund, August 1988.

Manville, A. M. 1988. Tracking plastic in the Pacific. *Defenders* (Nov./Dec.):10–15.

Margalef, R. 1968. *Perspectives in ecological theory.* University of Chicago Press.

Margalef, R., and M. Estrada. 1981. On upwelling, eutrophic lakes, the primitive biosphere, and biological membranes. In *Coastal and Estuarine Science, vol. 1: Coastal Upwelling*, F. A. Richards (Ed.), American Geophysical Union, Washington, D.C., pp. 522–529.

Massom, R. A. 1988. The biological significance of open water within the sea ice covers of the polar regions. *Endeavour* (New Series) 12:21–27.

May, R. M. 1988. How many species are there on Earth? *Science* 241:1441–1448.

McGibben, W. 1989. The end of nature. *The New Yorker*, September 11, pp. 47–105.

McGowan, J. A. 1986. The biogeography of pelagic ecosystems. In *Pelagic Biogeography: Proceedings of an International Conference, The Netherlands, 29 May–5 June, 1985*, Unesco Technical Papers in Marine Science, No. 49, pp. 191–200.

McGowan, J. A., and P. W. Walker. 1979. Structure in the copepod community of the North Pacific central gyre. *Ecological Monographs* 49:195–226.

McGowan, J. A., and P. W. Walker. 1985. Dominance and diversity maintenance in an oceanic ecosystem. *Ecological Monographs* 55:103–118.

McLusky, D. S. 1981. *The Estuarine Ecosystem.* John Wiley and Sons, New York.

McNeely, J. A. 1988. *Economics and biological diversity.* International Union for the Conservation for Nature and Natural Resources, Gland, Switzerland.

McNeely, J. A., K. R. Miller, W. V. Reid, R. A. Mittermeier, and T. B. Werner. 1990. Conserving the world's biological diversity. International Union for Conservation of Nature and Natural Resources, World Resources Institute, Conservation International, World Wildlife Fund–US, and the World Bank, Washington, D.C.

Monastersky, R. 1988. Plentiful plankton noticed at last. *Science News* 134:68.

Morse, D. E., and A. N. C. Morse. 1988. Chemical signals and molecular mechanisms: Learning from larvae. *Oceanus* 31:37–43.

Myers, N. 1979. *The sinking ark.* Pergamon Press, Oxford.

National Science Board. 1988. *The role of the National Science Foundation in polar regions.* NSB-87-128. National Science Foundation, Washington, D.C.

National Science Board. 1989. *Loss of biological diversity: A global crisis requiring international solutions.* NSB-89-171. National Science Foundation, Washington, D.C.

Olsen, S., L. Z. Hale, R. DuBois, D. Robadue, and G. Foer. 1989. *Integrated resources management for coastal environments in the Asia Near East Region.* International Coastal Resources Management Project, University of Rhode Island.

Pacific Science Association, Science Committee on Coral Reefs. 1988. *Coral Reef Newsletter,* No. 19.

Pain, S. 1988. No escape from the global greenhouse. *New Scientist* (November 12): pp. 38–43.

Paine, R. T. 1966. Food web complexity and species diversity. *American Naturalist* 100:65–75.

Paine, R. T. 1984. Some approaches to modeling multispecies systems. In *Exploitation of Marine Communities,* R. M. May (Ed.), Springer-Verlag, New York, pp. 191–207.

Paine, R. T., J. C. Castillo, and J. Cancino. 1985. Perturbation and recovery patterns of starfish-dominated intertidal assemblages in Chile, New Zealand and Washington State. *American Naturalist* 125:679–691.

Pianka, E. R. 1966. Latitudinal gradients in species diversity: A review of concepts. *American Naturalist* 100:33–46.

Pianka, E. R. 1988. *Evolutionary ecology.* Harper and Row, New York, pp. 309–314.

Pielou, E. C. 1979. *Biogeography.* Wiley-Interscience, New York.

Pimm, S. L. 1984. The complexity and stability of ecosystems. *Nature* 307:321–326.

Post, W. M., T. H. Peng, W. R. Emanuel, A. W. King, V. H. Dale and D. L. DeAngelis. 1990. The global carbon cycle. *American Scientist* 78:310–326.

Ray, G. C. 1988. Ecological diversity in coastal zones and oceans. Chapter 4 in *Biodiversity,* E. O. Wilson (Ed.), National Academy Press, Washington, D.C., pp. 36–50.

Reid, W. V., and K. R. Miller. 1989. *Keeping options alive: The scientific basis for conserving biodiversity.* World Resources Institute, Washington, D.C.

Bibliography

Rex, M. A. 1981. Community structure in deep-sea benthos. *Annual Review of Ecology and Systematics* 12:331–353.

Roughgarden, J., S. Gaines, and H. Possingham. 1988. Recruitment dynamics in complex life cycles. *Science* 241:1460–1466.

Rowe, G. T. 1981. In *Coastal and Estuarine Science, vol. 1: Coastal Upwelling*, F. A. Richards (Ed.), American Geophysical Union, Washington, D.C., pp. 464–471.

Rudyakov, Y. A. 1987. Ecosystems of coastal waters as a component of the biological structure of the ocean. *Oceanology* 27:479–481.

Ruggieri, G. D. 1976. Drugs from the sea. *Science* 194:491–497.

Rygg, B. 1985. Distribution of species along pollution-induced diversity gradients in benthic communities in Norwegian fjords. *Marine Pollution Bulletin* 16:469–474.

Saenger, P., E. J. Hegerl, and J. D. S. Davie (Eds.). 1983. Global status of mangrove ecosystems. *The Environmentalist*, vol. 3, supplement 3.

Sale, P. F. 1980. The ecology of fishes on coral reefs. *Oceanography and Marine Biology, Annual Review* 18:367–421.

Salm, R. V., with J. R. Clark. 1984. *Marine and coastal protected areas: A guide for planners and managers.* International Union for the Conservation of Nature and Natural Resources, Gland, Switzerland.

Sanders, H. L. 1968. Marine benthic diversity: A comparative study. *American Naturalist* 102:243–282.

Santelices, B. 1990. Patterns of reproduction, dispersal and recruitment in seaweeds. *Oceanography and Marine Biology Annual Review* 28:177–276.

Schopf, T. J. M., J. B. Fisher, and C. A. F. Smith III. 1978. Is the marine latitudinal diversity gradient merely another example of the species area curve? In *Marine Organisms: Genetics, Ecology, and Evolution*, Battaglia and Beardmore (Eds.), Plenum Press, New York, pp. 365–386.

Sherman, K. 1986. Introduction to parts one and two: Large marine ecosystems as tractable entities for measurement and management. Chapter 1 in *Variability and Management of*

Large Marine Ecosystems, K. Sherman and L. Alexander (Eds.), AAAS Selected Symposium 99:3–8.

Sherman, K., J. R. Green, J. R. Goulet, and L. Ejsymont. 1983. Coherence in zooplankton of a large northwest Atlantic ecosystem. *Fishery Bulletin* 81:855–862.

Sherman, K., Grosslein, D. Mountain, D. Busch, J. O'Reilly, and R. Theroux. 1988. The continental shelf ecosystem off the northeast coast of the United States. In *Ecosystems of the World,* 27: *Continental Shelves,* H. Postma and J. J. Zijlstra (Eds.), Elsevier, Amsterdam, pp. 279–337.

Sherr, E. B. 1989. And now, small is plentiful. *Nature* 340:429.

Sissenwine, M. P. 1986. Perturbation of a predator-controlled continental shelf ecosystem. Chapter 5 in *Variability and Management of Large Marine Ecosystems,* K. Sherman and L. Alexander (Eds.), AAAS Selected Symposium 99:55–85.

Smith, P. J., and Y. Fugio. 1982. Genetic variation in marine teleosts: High variability in habitat specialists and low variability in habitat generalists. *Marine Biology* 69:7–20.

Smith, V. K. 1988. Resource evaluation at the crossroads. *Resources* 90:2–6.

Sousa, W. P. 1984. The role of disturbance in natural communities. *Annual Review of Ecology and Systematics* 15:353–391.

Spight, T. T. 1977. Diversity of shallow-water gastropod communities on temperate and tropical beaches. *American Naturalist* 982:1077–1097.

Steele, J. H. 1985. A comparison of terrestrial and marine ecological systems. *Nature* 313:355–358.

Steele, J. H., S. Carpenter, J. Cohen, P. Dayton, R. Ricklefs. 1989. *Comparison of terrestrial and marine ecological systems: Report of a workshop held in Santa Fe, New Mexico.* Prepared by the Steering Committee.

Stegeman, J. J., P. J. Kloepper-Sams, and J. W. Farrington. 1986. Monooxygenase induction and chlorobiphenyls in the deepsea fish *Coryphaenoides Armatus. Science* 231:1287–1289.

Stehli, F. G., and J. W. Wells. 1971. Diversity and patterns in hermatypic corals. *Systematic Zoology* 20:115–126.

Stehli, F. G., R. G. Douglas, and N. D. Newell. 1969. Generation and maintenance of gradients in taxonomic diversity. *Science* 164:947–949.

Tiner, R. 1984. *Wetlands of the United States: current status and recent trends.* U.S. Department of Interior, Washington, D.C., 57 pp.

Toggweiler, J. R. 1989. Are rising and falling particles microbial elevators? *Nature* 337:691–692.

Turner, S. M., et al. 1988. The seasonal variation of dimethyl sulfide and dimethylsulfoniopropionate concentrations in nearshore waters. *Limnology and Oceanography* 33:364–375.

Underwood, A. J., and E. J. Denley. 1984. Paradigms, explanations, and generalizations in models for the structure of intertidal communities on rocky shores. Chapter 11 in *Ecological Communities: Conceptual Issues and the Evidence*, D. R. Strong, Jr. (Ed.), Princeton University Press, Princeton, N.J., pp. 151–180.

United Nations Environment Programme. 1982. *Achievements and planned development of UNEP's Regional Seas Programme and comparable programmes sponsored by other bodies.* UNEP Regional Seas Reports and Studies, No. 1. UNEP, Nairobi.

United Nations Environment Programme. 1988. *1987 Annual report of the executive director.* UNEP, Nairobi.

U.S. Agency for International Development. 1985. *U.S. strategy on the conservation of biological diversity: An interagency task force report to Congress.* U.S. AID, Washington, D.C.

U.S. Agency for International Development. 1988. *Progress in conserving tropical forests and biological diversity in developing countries: The 1987 annual report to Congress on the implementation of Sections 118 and 119 of the Foreign Assistance Act, as amended.* U.S. AID, Washington, D.C.

U.S. Congress, Committee on Merchant Marine and Fisheries. 1988. *Coastal waters in jeopardy: Reversing the decline and protecting America's coastal resources.* Oversight Report of the Subcommittee on Fisheries and Wildlife Conservation and the Environment and the Subcommittee on Oceanography of the Committee on Merchant Marine and Fisheries. U.S. Government Printing Office, Washington, D.C.

U.S. Congress, House, Committee on Merchant Marine and Fisheries. 1988. *Report on the National Marine Sanctuaries Authorization Act of 1988*. Rept. 100–739, Part 1 to accompany H.R. 4208. 100th Cong., 2d Session. U.S. Government Printing Office, Washington, D.C.

U.S. Congress, Office of Technology Assessment. 1987. *Technologies to maintain biological diversity*. OTA-F-330. U.S. Government Printing Office, Washington, D.C.

U.S. Congress, Office of Technology Assessment. 1987. *Wastes in marine environments*. OTA-O-334. U.S. Government Printing Office, Washington, D.C.

U.S. Department of Commerce, National Oceanic and Atmospheric Administration, National Marine Pollution Program Office. 1986. *Summary of federal programs and projects: FY 1985 update*. NOAA, Washington, D.C.

U.S. Department of Commerce, National Oceanic and Atmospheric Administration. 1986. *NOAA fishery management study*. NOAA, Washington, D.C.

U.S. Department of Commerce, National Oceanic and Atmospheric Administration, National Marine Fisheries Service. 1987. *Marine Mammal Protection Act of 1972, annual report: 1986/87*. U.S. Department of Commerce, Washington, D.C.

U.S. Department of Commerce, National Oceanic and Atmospheric Administration. 1988. *National Estuarine Reserve Research System, status report*. Marine and Estuarine Management Division/OCRM/NOS/NOAA, Washington, D.C.

U.S. Department of Commerce, National Oceanic and Atmospheric Administration. 1988. *National Marine Sanctuary Program: status report*. Marine and Estuarine Management Division/OCRM/NOS/NOAA, Washington, D.C.

U.S. Department of the Interior, U.S. Fish and Wildlife Service. 1988. *Management of the national wildlife refuges, draft environmental impact statement*. Department of the Interior, Washington, D.C.

U.S. Environmental Protection Agency, Office of Marine and Estuarine Protection. 1986. Near coastal waters: Strategic options paper. Unpublished report.

Venrick, E. L. 1982. Phytoplankton in an oligotrophic ocean: Observations and questions. *Ecological Monographs* 52:129–154.

Venrick, E. L. 1990. Phytoplankton in an oligotrophic ocean: species structure and interannual variability. *Ecology* 71: 1547–1563.

Venrick, E. L., J. A. McGowan, D. R. Cayan, T. L Hayward. 1987. Climate and chlorophyll a: Long-term trends in the central north Pacific Ocean. *Science* 238:70–72.

Vermeij, G. J. 1978. *Biogeography and adaptation: Patterns of marine life.* Harvard University Press, Cambridge, Mass.

Warwick, R. M., and Ruswahyuni. 1987. Comparative study of the structure of some tropical and temperate marine soft-bottom macrobenthic communities. *Marine Biology* 95:641–649.

Wells, J. W. 1957. Coral reefs. Chapter 20 in *Treatise on Marine Ecology and Paleoecology,* J. W. Hedgepeth (Ed.), Memoir 67, vol. 1, Geological Society of America, New York, pp. 609–631.

Wells, S. M., and J. G. Barzdo. In press. International trade in marine species—is CITES a useful control mechanism? *Journal of Coastal Management.*

Wiegert, R. G., and L. R. Pomeroy. 1981. The salt-marsh ecosystem: A synthesis. Chapter 10 in *The Ecology of a Salt Marsh,* R. Pomeroy and R. G. Wiegert (Eds.), Springer-Verlag, New York, pp. 218–239.

Wilson, E. O. (Ed.). 1988. *Biodiversity.* National Academy Press, Washington, D.C.

World Commission on Environment and Development. 1987. *Our common future.* Oxford University Press, Oxford.

World Resources Institute, International Institute for Environment and Development. 1986. *World resources 1986.* Basic Books, Inc., New York.

World Resources Institute, International Institute for Environment and Development. 1987. *World resources 1987.* Basic Books, Inc., New York.

World Resources Institute, International Institute for Environment and Development. 1988. *World resources 1988–89.* Basic Books, New York.

About Friends of the Earth and the Oceanic Society

IN early 1989 three national environmental organizations—Friends of the Earth, the Environmental Policy Institute, and the Oceanic Society—joined forces to create a new organization capable of supporting citizen action to protect the environment on a global scale. These organizations were drawn into the merger by a realization that the global environment is deteriorating and that new efforts are needed to address these problems directly. The mission of the newly organized Friends of the Earth, U.S., is to create an independent, global environmental advocacy organization that will work at local, national, and international levels to protect the planet from environmental disaster; preserve biological, cultural, and ethnic diversity; and involve citizens in decisions affecting their environment and their lives.

Friends of the Earth, U.S., is part of an international network of environmental advocacy organizations known as Friends of the Earth International. It is one of the most diverse international advocacy organizations in the world, with representatives in forty countries around the globe. Affiliates are located in the northern and the southern hemispheres, in highly developed and in less well developed worlds, in small and large countries. They are as widespread as Australia, Hong Kong, Estonia, Ghana, the Netherlands, Canada, and Brazil. Although each group has its own priorities and concerns, several international campaigns are coordinated through the international network: protecting the ozone layer, saving tropical forests, improving East-West communication and cooperation on environmental issues, and preserving oceans and marine life.

Friends of the Earth, U.S., is actively involved in the international marine campaign and in numerous local, national, and international issues related to ocean protection. The Oceanic Society brought to the organization a highly respected reputation based on two decades of work on ocean issues. Prior to the 1989 merger, the Oceanic Society was dedicated solely to protecting the oceans for the people and the wildlife that depend on them for life, livelihood, and enjoyment. Through education, research, and conservation, the Oceanic Society has promoted understanding and stewardship of marine and coastal environments. The organization traditionally has advocated marine policies, practices, and uses that establish a nondestructive relationship between human and marine life. Now, as a project of Friends of the Earth, the Oceanic Society continues the tradition, serving as a voice for the silent seas and their inhabitants, a voice for conservation of marine biodiversity.

About the Authors

BOYCE THORNE-MILLER is a marine biologist with the Oceanic Society Project of Friends of the Earth, U.S., where she works to involve science and scientists in the marine environmental policy-making process at international, national, and grassroots levels. Her range of environmental work has included the protection of marine biological diversity through the international regulation of hazardous waste disposal in marine waters, national regulation of the disposal of contaminated dredge materials and sewage sludge in coastal waters, identification of biological indicators of marine pollution, and the designation of marine sanctuaries. She received an M.S. in oceanography from the University of Rhode Island, where she later worked as a research associate. She was involved in field research on phytoplankton and seaweeds in Rhode Island estuaries and also in mariculture research. She has taught several college-level biology courses and continues to incorporate education, in a less formal way, into her environmental work. She and her family live in rural Maryland, where they and their neighbors are continually fighting to prevent the desecration of their own "backyard" environment.

JOHN G. CATENA is currently a policy analyst for the State of Maine Coastal Management Program, where he is specializing in marine policy and regional cooperation in the Gulf of Maine. He previously worked for several years as a policy analyst for the Oceanic Society, specializing in national and international marine environmental policy. Mr. Catena has worked on a variety of marine environmental issues, including ocean dumping, seabed mining, and coastal water quality and habitat protection. He holds a B.S. in marine science from the University of South Carolina and an M.A. in marine affairs from the University of Rhode Island. His scuba diving experiences have allowed him a firsthand look at the biological diversity of numerous coastal habitats, including Narragansett Bay, the Florida Keys, Hawaii, the Caribbean, and the Mediterranean. John and his wife, Jessica, live in Maine.

170

Index

Abalone, 34, 35
Abatement of marine pollution,
  74–78
  coastal marine degradation and,
    75, 76
  direct regulation of discharges,
    77–78
  intentional uses of oceans for dis-
    posal of wastes, 75, 76
  international action, 78
  laws to regulate, 76–78
  open-ocean ecosystems and, 75
  phasing out of pollution, 75, 76
  policies for, 75–76
  regulation of discharges, 75–76
  water quality standards, 77
Adaptive management, 79–80
Algae, 13, 34, 48, 56
  calcareous, 50–51, 52
American Àssociation for the Advancement of Science, 121
Anchovy depletion, 20
Animals:
  in estuaries and wetlands, 49–50
  on rafts of *Sargassum*, 67
  reestablishment of, 84–85
  regulation of hunting of, 83–84
  vertical zones of, 33
Antarctic, 70–71
  laws to protect, 113–14
Aquariums, 85
Arctic, 70–71
Arctic National Wildlife Refuge, 71,
  93–94, 106
Athens Protocol, 78
Atlantic salmon, 85
Atmospheric pollutants, 17–18

Bacteria, 28, 34, 38
  in the open ocean, 65, 69–70
Barcelona Convention, 78
Benthic communities, 28, 30
  classifications of, 39
  coastal, *see* Coastal benthic
    ecosystems
  coral reefs, *see* Coral reefs

deep sea, *see* Deep-sea benthic
  ecosystems
depth and, 37, 60
keystone species and, 24–25
subtidal continental shelf, 53–55
Bioaccumulation, 18–19
Biochemistry, 34–35
Biological diversity, *see* Marine bio-
  logical diversity
Biomagnification, 19
Biomes, 40
  coastal, *see* Coastal marine
    ecosystems
  oceanic, *see* Oceanic marine
    ecosystems
Biosphere Reserve Program, 115
"Black list," 108
Bush, George, 105

California Cooperative Oceanic Fish-
  eries Investigations (CalCOFI),
    118–19
Capitella, 27
Carbon dioxide, 11, 21, 71
Cartagena Convention, 111
Channel Islands NMS, 90
Characteristic diversity, 8–9, 24
Chemosynthesis, 63–64
Chukchi Sea, 71, 106
Clean Water Act (CWA), 97–99
Coastal Barrier Resources Act, 101
Coastal Barrier Resources System
  (CBRS), 101
Coastal basins, 58
Coastal habitats, 16
Coastal marine ecosystems, 41–58
  biological dispersal, 42
  boundaries of, 42
  coastal basins, 58
  coastal benthic, *see* Coral reefs;
    Estuaries and wetlands;
    Rocky intertidal and subtidal
    shores; Sandy shores and
    mud flats; Subtidal continen-
    tal shelf

171

Also Available from Island Press

Ancient Forests of the Pacific Northwest
By Elliott A. Norse

Balancing on the Brink of Extinction: The Endangered Species Act and Lessons for the Future
Edited by Kathryn A. Kohm

Better Trout Habitat: A Guide to Stream Restoration and Management
By Christopher J. Hunter

The Challenge of Global Warming
Edited by Dean Edwin Abrahamson

Coastal Alert: Ecosystems, Energy, and Offshore Oil Drilling
By Dwight Holing

The Complete Guide to Environmental Careers
The CEIP Fund

Economics of Protected Areas
By John A. Dixon and Paul B. Sherman

Environmental Agenda for the Future
Edited by Robert Cahn

Environmental Disputes: Community Involvement in Conflict Resolution
By James E. Crowfoot and Julia M. Wondolleck

Fighting Toxics: A Manual for Protecting Your Family, Community, and Workplace
Edited by Gary Cohen and John O'Connor

Hazardous Waste from Small Quantity Generators
By Seymour I. Schwartz and Wendy B. Pratt

Holistic Resource Management Workbook
By Alan Savory

In Praise of Nature
Edited and with essays by Stephanie Mills

The New York Environment Book
By Eric A. Goldstein and Mark A. Izeman

Overtapped Oasis: Reform or Revolution for Western Water
By Marc Reisner and Sarah Bates

Permaculture: A Practical Guide for a Sustainable Future
By Bill Mollison

Plastics: America's Packaging Dilemma
By Nancy A. Wolf and Ellen D. Feldman

The Poisoned Well: New Strategies for Groundwater Protection
Edited by Eric Jorgensen

*Race to Save the Tropics: Ecology and Economics for a
Sustainable Future*
Edited by Robert Goodland

Recycling and Incineration: Evaluating the Choices
By Richard A. Denison and John Ruston

The Rising Tide: Global Warming and World Sea Levels
By Lynne T. Edgerton

Rush to Burn: Solving America's Garbage Crisis?
From *Newsday*

Saving the Tropical Forests
By Judith Gradwohl and Russell Greenberg

War on Waste: Can America Win Its Battle With Garbage?
By Louis Blumberg and Robert Gottlieb

Western Water Made Simple
From *High Country News*

Wetland Creation and Restoration: The Status of the Science
Edited by Mary E. Kentula and Jon A. Kusler

Wildlife and Habitats in Managed Landscapes
Edited by Jon E. Rodiek and Eric G. Bolen

For a complete catalog of Island Press publications, please write:
Island Press, Box 7, Covelo, CA 95428.
Or call 1-800-828-1302.

Island Press
Board of Directors

PETER R. STEIN, CHAIR
Managing Partner, Lyme Timber Company
Board Member, Land Trust Alliance

DRUMMOND PIKE, SECRETARY
Executive Director
The Tides Foundation

ROBERT E. BAENSCH
Director of Publishing
American Institute of Physics

PETER R. BORRELLI
Editor, *The Amicus Journal*
Natural Resources Defense Council

CATHERINE M. CONOVER

GEORGE T. FRAMPTON, JR.
President
The Wilderness Society

PAIGE K. MACDONALD
Executive Vice President/
Chief Operating Officer
World Wildlife Fund/The Conservation Foundation

HENRY REATH
President
Collectors Reprints, Inc.

CHARLES C. SAVITT
President
Center for Resource Ecomonics/Island Press

SUSAN E. SECHLER
Director
Rural Economic Policy Program
Aspen Institute for Humanistic Studies

RICHARD TRUDELL
Executive Director
American Indian Lawyer Training Program

2816